The Fourth Generation

The Fourth Generation

An Eco-thriller

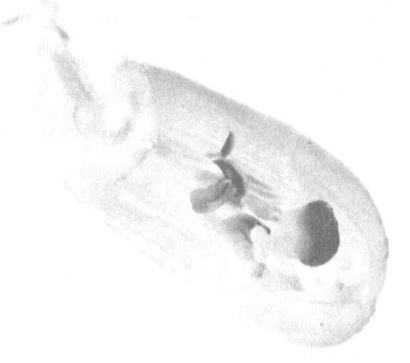

Roy Mankovitz

and

Alan Mankovitz

Weeping Willow Books

The Fourth Generation is a work of fiction. No events, or persons depicted here are real, and no inference should be made to actual events.

Book design and cover image by Don Mitchell

ISBN 978-0-9990994-9-0

Weeping Willow Books
Santa Barbara, CA

To the memory of Roy Jack Mankovitz

To my late husband, Roy Jack Mankovitz: You were a force of nature and a man born before his time. I am so proud of the extensive research you conducted and the deep-seated passion you poured into the development of this exciting eco-thriller. Your love and respect for Mother Nature was always your guiding force—and into her arms you returned so unexpectedly.

To my stepson Alan Mankovitz: You carried the work of your father forward in a respectful and loyal fashion. Thank you for your creative brilliance and dedication to your father's ideas for this book.

—Kathleen A. Barry, Ph.D.

I dedicate this book to my hero. Some people never get to meet their hero, but I got to grow up with mine. This is for you, Dad.

—Alan Mankovitz

CONTENTS

ONE

March 2030

RANDY HALL AWOKE IN DARKNESS, as he always did. His clock read 3:47. Ordinarily, he'd have rolled out of bed and into the shower, ready to start his day. But this morning he lay there, staring at what he could make out of the ceiling.

By now, he imagined, Walter Conroy had reviewed not only the notes Randy had zapped him but his own, from the tests he'd performed years ago. Walter would have seen that Veregro couldn't continue with business as usual in light of such conclusive evidence. Maybe he'd already zapped Randy back.

The thought was enough to propel Randy out of bed and into his small galley kitchen to check his tablet. But the only messages were his morning newsfeed from the *Times* and the usual array of offers to enlarge his penis and deposit a million Euros into his bank account as soon as he provided the sender with the number. Annoyed, Randy switched off the screen, and started a pot of coffee.

He was surprised Walter hadn't been in touch. But maybe Walter had decided zapping him wasn't a good idea. Maybe Veregro kept tabs on employees' messages. If so, Walter would know it. But wouldn't the company want to know what they'd found? Randy felt the acid taste of his first sip of coffee without even sipping the coffee itself.

Something wasn't right.

Randy switched the screen back on and logged into his office mail. Or rather, he tried to log into his office mail. His password wasn't working. Uh-oh. Did he make a mistake trusting Walter?

He got up and peered out from behind the closed blinds of the kitchen window to see fresh snow falling through the halo of the sodium light that illuminated the apartment complex's parking lot. The parked cars' windshields were all covered. Nobody was out there. He let the blind snap itself shut.

Randy wasn't used to uneasy feelings. He intentionally kept his world ordered, his routine unvarying. Now he opened the cupboard that contained the plates instead of the coffee cups, and, after he'd opened the right cupboard, he dropped the cup to the floor, where it shattered.

Shaking, Randy got down another cup and carefully poured from the carafe. Then he added his usual cream and sugar, and stirred five times. The routine calmed him a bit and he got down a bowl for his cornflakes.

While he ate, Randy read the *Times* newsfeed, sometimes clicking through to watch a video. He clicked through on a headline that read, "MODS

Reaching Plague Proportions," and a blonde talking head hologram emerged from his screen.

"Incidences of Multiple Organ Dysfunction Syndrome have now been reported in cities across the country, with deaths confirmed in St. Louis, Boise, Gainesville, and Albany," she recited solemnly. Holograms of a series of hospitals flashed in quick succession behind her before being replaced by a large-fonted graphic that read, "The New Plague?"

"Hospital officials in those cities would neither confirm nor deny whether the deaths were connected to MODS, and cautioned against a panic, despite the growing number of unexplained deaths," the talking head continued. "Please click through to read the symptoms, and click back to the *Times* throughout the day for updates."

Ordinarily, by this time, Randy would already be getting ready to head to the lab, but something nagged at him. He poured and doctored another cup of coffee, stirred it five times, then opened a search window and typed in "symptoms MODS." A million hits. He clicked through on the first. It wasn't a Med-Site, but Randy read on anyway. He was a scientist. He could sort the BS from the truth.

MODS seems to begin with an inability to digest food, possibly because the intestine begins to feed on itself. Concurrently, or sometimes subsequently, a rash appears on the belly or lower arms, typically raised welts that resemble hives in both appearance and itchiness. This rash spreads rapidly, so that often within an hour the entire body is quickly covered in these itching oozing sores. Finally (and often as a great relief), sudden dilations of the heart begin to occur, the last resulting in death.

Holy fucking shit. Randy didn't check the time. He opened a line to Walter Conroy's tablet number and clicked "call."

While he ate breakfast, Walter Conroy watched the news on his tablet. North Korean leader threatening to nuke South Korea out of the China Sea. Another tsunami in the South Pacific. Deaths from MODS reported in Houston, Chicago, Los Angeles, and Denver, as well as Toronto, Barcelona, Sydney, and St. Petersburg. Jesus. It wasn't just a disease. It was an epidemic.

The blonde talking head reappeared on his screen. "There is a glimmer of hope, though. Our San Francisco correspondent, Simon Keller, has discovered a secret valley where, not only is no one sick, but women are having babies who are healthy and strong. Simon?"

Walter pulled the tablet closer and adjusted the volume. A scrawny young guy who looked as if he'd been promoted from blogger the day before twitched in front of a handheld. "Hi Connie," he said, his voice cracking. Walter barked a laugh.

"What can you tell us about the Orchard, Simon?"

As the kid began to talk, his image was replaced by an aerial shot of swaths of fruit trees, tucked into the foothills of what looked like the California coastal range. "The Orchard was founded twenty-five years ago by Charles Winters." A snap of a distinguished middle-aged man appeared briefly before the flyover resumed. "After studying ethnobiology and primatology for a number of years, Winters purchased a remote valley in western Marin County and moved his family there."

Another snap, of the same man, this time with a wife and young daughter.

"Before long, a number of veterans of the third Gulf war had moved there, too. They're quiet—their neighbors in Bolinas say they come into town for supplies occasionally, and that there's a trader, whom I met, who comes to sell the fruits and meat they produce, but, for the most part, they keep to themselves."

The camera began another aerial sweep of the lush, green valley, and Walter got up to pour himself more coffee without turning the story off. When he turned back, he was struck by a memory that hit him like two strong arms on his shoulders, pushing him down to the couch.

On the TV was a still image of a pretty young girl; autumn-colored hair and sparkling green eyes. Walter had seen this picture before, on a different newscast, eighteen years ago. He stared at the image until it disappeared and the aerial pan resumed, then reached out, trance-like, to turn up the volume.

"Charlotte Winters returned to the commune after her infamous association with the Angell Farm, eighteen years ago. John Angell—" and there he was, the long-dead farmer Walter wished he'd never met, "killed his family and himself because Veregro, in the person of its corporate counsel, Charlotte Winters, had been demanding he pay licensing fees for using Veregro seed."

The photo-over was now of the Angell Farm, eighteen years ago. Jesus. Where did they come up with all this stuff? The bargain basement? "At that time," the kid was saying, his voice growing more assured with each sentence, "Veregro had been seeking licensing fees from any farmer whose product tested positive for Veregro seed, whether or not they'd purchased the seed themselves."

"John Angell—" the background cut back to the farmer, "refused. He killed himself—and his family—rather than pay what he called blood money."

Now the picture of Charlotte was back. Walter had been about to take a sip of coffee but instead set the cup down, shakily. "Charlotte Winters was found to have been the Veregro employee applying pressure to John Angell, and was subsequently fired, but was never charged with a crime. After congressional hearings, Veregro was cleared of any wrongdoing."

The reporter reappeared on the screen, standing on the wooden sidewalk of one of those coastal California towns that hadn't changed since 1965—well, the ones that hadn't been washed out by the tsunamis, anyway.

"I came to Bolinas, California, after hearing reports that people on Charlotte Winters' Orchard aren't getting sick. What, I wondered, does Charlotte Winters know that the rest of us don't?

"I followed the Orchard's trader back to the Orchard last week in an attempt to ask Ms. Winters herself. Unfortunately, I was unceremoniously escorted from the premises.

"So instead, I've been asking around town, to find out what the people here can tell me about Ms.

Winters and her mysterious Orchard." He turned to beckon someone into the frame, and a middle-aged woman appeared. Sun-aged skin, sun-bleached hair. Taller than the kid.

"Mattie Kallen works at the Mercantile, where Orchies sometimes come to buy tools and supplies." The reporter turned halfway toward the woman and extended a hand mic toward her. "Mattie, why do you think no one is sick at the Orchard?"

The woman leaned toward the mic, but faced the handheld, as most people did, instinctively. "I can't say I know why. But they live natural lives over there. Grow their own food. And, you know, there's all kinds of things they don't or won't eat. They eat meat, but they raise it themselves. Won't go near eggs or wheat or lots of other things I can't even remember. They're hardly ever sick. My son's friends with some of their kids, and they're good-looking kids. Strong." She smiled, then shrugged. She knew her fifteen seconds were over.

The reporter nodded and the handheld cut to him alone. "While no one seems to know what's causing MODS, Charlotte Winters's Orchard remains disease-free. What does Charlotte Winters know? And why won't she share it?"

The kid paused for a dramatic beat. Then he closed. "Reporting from Bolinas, in far western Marin County, California, I'm Simon Keller."

The talking head came back on-screen. "We'll be following this story as it develops," she said, "and you can read more details at Simon's blog at SimonSays. com. Thank you, Simon." The station cut to a Chevy commercial and Walter muted the sound.

The kitchen was quiet. Charlotte Winters. Grow their own food. Won't go near eggs or wheat or lots of other things. Stick to themselves. No one was sick. No one mysteriously dead. Healthy babies being born. What did Charlotte Winters know?

It was not yet noon but Walter was fatigued. His head hurt from hours spent reading Randy's report and then more hours comparing it to his own report. Now Charlie Winters is on the television. Is this a dream, a perfect circle, or the Bermuda Triangle? It all crossed Walter's mind.

Even though it had been light for hours, Randy hadn't opened the blinds. Periodically, he sidestepped to the window and snuck a look outside.

The snow had stopped and the parking lot was tracked with the footprints of neighbors who'd walked to their cars, and the tire tracks of the cars as they'd left. The lot was mostly empty, Randy's old Honda sitting lonely at its heart.

As he watched, a car pulled up next to his. An Escalade. Black. Randy had watched enough crime videos to know that the only people who drove black Escalades were either very, very good or very, very bad.

He had a feeling this wasn't one of the good guys.

Randy went into his closet and found the binoculars his father had given him after he'd made Eagle Scout. He set his glasses on the kitchen table and then adjusted the lenses to bring the license plate into focus. It wasn't a Nebraska plate, looked like Kansas or Missouri. No. Illinois. He could see it now.

Ducking down, he typed the license number into a new message addressed to Walter Conroy's private address. Then he opened the encrypted file he'd uploaded to the Cloud and appended its contents to the same message.

He edged back to the window and looked out again. The Escalade was gone. Randy used the binoculars to track its path through the snow. It disappeared around the back of his building.

He heard the door from the stairs open onto his hallway. He hit "send" and watched the screen until the message had been successfully delivered. Then he switched to his main drive and clicked "Reformat."

Are you sure? the screen asked.

Yes! Yes, he was sure. Randy typed in the password.

This operation will erase all files on your hard drive. Have you backed up?

Footsteps came along the hall, then stopped outside Randy's door. Yes, I've backed up, Randy clicked.

There was a light knock on the door. Randy froze and silently stared. He heard a few more taps, then the dead bolt slowly turned as if unlocked by an invisible force. The door knob turned and the door swung open.

A tall man, his mouth and nose covered by a gauze mask, his hands encased in latex lab gloves, slipped some tools into his coat pocket and pointed a gun at Randy. The man stepped into the apartment and calmly closed the door behind him.

"You may as well shoot me now. I'm not going to tell you anything," Randy said.

A deep chuckle came from behind the mask as the man stepped forward and pushed Randy down into a chair. "You think I'm here to get you to talk?" The man pulled a syringe from his pocket and stuck the needle into Randy's neck. Randy began to blink wildly as the liquid in the syringe emptied into his body. He stole one last look at his screen. *Re-format complete,* it read.

Randy smiled as his wild blinking slowed. And then stopped.

Two

One Day Earlier

Daisy was dead.

Randy knew it as soon as the fluorescent lights buzzed to life overhead. Most early mornings when he entered the lab, he heard the sound of Daisy on her treadmill even before he'd switched on the lights. This morning, though, there was only silence.

Randy hurried across the room without removing his down jacket. In the big cage tucked on the wide shelf directly above his desk, Huck huddled in a corner. He gave Randy a mournful look.

Don't anthropomorphize the lab rats, Randy reminded himself. As if he had any choice.

In the far corner opposite Huck, Daisy lay flat on her back, pretty pink paws daintily extended upward as if in a last plea for salvation. "Aw, Daisy," Randy said, keeping his voice low even though he was always the only one in the lab this early.

He unlatched the cage and first scooped out Huck, taking a moment to nuzzle his whiskers before setting him into the empty cage next door. He set in fresh pellets and water, and, after a last farewell look at Daisy, Huck began eating.

Randy turned his attention to Daisy. She'd been his favorite of all the rats. That was why he'd chosen to give her a shot at insemination, to keep her line going. And she'd loved being pregnant. Randy could tell.

Randy removed her body from the cage and cradled her briefly before getting down to what came next. He'd need blood and tissue samples. He'd need to perform an autopsy, recording every step. He wanted to get as much done as he could before everyone else arrived.

He looked at the big clock by the door. 5:30 a.m. He had two-and-a-half hours. He shucked his jacket, made a pot of coffee, and got to work.

Once Daisy's blood samples were whirring in the centrifuge, Randy awakened his computer. Randy liked to perform one task at a time, giving it his full attention. Pauline Wozeck, the perennially cheerful blonde at the next desk, claimed she could do a routine file purge with one hand and prepare lab samples with the other.

But lab work wasn't about multi-tasking. It was about taking one meticulous step at a time, documenting that step, and then moving on to the next one. Like most people, Pauline Wozeck got on Randy's nerves.

Randy, who'd grown up in Omaha, still lived in the apartment just north of UNO he'd moved into after college. He was an early riser—up by 4 a.m.

every morning—and he never tired of having the streets between his apartment and the Veregro lab, next to the river between the airport and downtown, to himself. This morning, there had been a fresh powdering of snow to cover the graying mounds that the plows had pushed aside all winter. As it was April, this might be the last of it until fall. Randy liked the snow. It was quiet. He'd be sorry to see it go.

Randy had been with Veregro almost fifteen years. He'd been hired before graduating college. He was already considered a gifted scientist who overlooked no detail.

That was why, when he'd discovered Walter Conroy's lab notes during a routine purge, he'd read them. The Veregro brass required routine file purges every six months, to keep the server clean, they said. But, unlike Pauline Wozeck, Randy couldn't simply highlight a block of files and hit the delete button. The research that had gone on before he and Pauline had been hired wasn't just interesting; sometimes it shed light on precisely what they were doing right this minute.

Walter Conroy was now Veregro's vice president for communications, but unlike the rest of the Suits, he sometimes still came down to the lab, back to his origins. Once, when Randy had been about to inject one of Daisy's predecessors with a new product, Walter held out a hand and asked if he could do it. Randy had watched with growing approval as Walter had first washed his hands, next donned fresh gloves from the dispenser, then carefully swabbed the site with alcohol before executing such a meticulous injection that the rat barely registered its sting.

So it was only natural that, when Randy had seen Walter's initials appended to a file that was scheduled to be purged, he had to open it. But then it turned out the file was encrypted. Randy's computer skills were largely outdated even though he'd taken the requisite number of computer science courses at UNO. Still, they'd turned out to be good enough to crack the code on an old server like the one Veregro continued to use.

Randy had to read the notes twice to be certain he was seeing what he thought he was. How could Veregro continue to make, let alone sell, both genetically modified seed and the insect repellent Veresate, if this was the end result? And how could Walter Conroy have buried this?

Sure, Walter had tested for only a few years, with only a few generations. But with results as conclusive as these—infertility, shrunken testicles in males and ovarian cysts in females, organ failure, slow and painful death—more had been clearly called for. Except the file was buried in 2012. And encrypted. And GM seed and Veresate were Veregro's biggest sellers.

On his own, Randy had set out to replicate Walter's tests. Everyone knew Randy got to the lab before sunrise, and he took advantage of those early hours, when no one else was around. Randy was a company man. He'd set out to prove that Walter Conroy's tests had been flawed.

Except now, Daisy was dead.

Randy glanced at the centrifuge. There were still a few minutes left on the timer, so he typed in the password of the encrypted file. Then he tapped a few keys and opened a new encrypted site in the Cloud, and uploaded everything. Company man or no, Randy was going to have to share his findings.

He looked at the clock. It was too early to put in a call to Walter Conroy. Randy would complete his tests and call him then. If they corroborated Walter's, the shit was going to hit the fan. Walter Conroy would know what to do. After all, he was the one who'd started it.

Walter Conroy's hand strayed to the other side of the bed, only to find Kylie wasn't there. Kylie never got up before he did. Then Walter remembered. Kylie was dead. He'd buried her yesterday.

Fresh tears rose in Walter's eyes, and he hadn't even opened them yet. How was he going to get through the day, let alone the rest of his life? He and Kylie had been married for almost eighteen years. This pregnancy had seemed as if, unlike the ones that had preceded it, it was going to come to term. Then Kylie had broken out in that rash, and they'd both known she had MODS. At the end, one short week later, Walter had begged the doctor to up her morphine. He couldn't bear seeing her suffer. The doctor had shown him how to work the pump, and then had left the room.

From the kitchen came the sound of someone grinding coffee beans, and Walter was momentarily confused. Maybe he'd dreamed it. Maybe Kylie wasn't dead. He kept his eyes shut, as if doing so would avert the inevitable. Then the house phone rang, and he heard his sister Rachel answer. The house phone hadn't rung in years. This week, it had hardly stopped.

"I don't think he's up yet," Rachel said, then waited. "I can't imagine what could possibly be important enough to wake him. Are you aware his wife died? Are you aware we just buried her yesterday?"

Walter opened his eyes to the empty pillow beside his.

"Why don't you call his secretary?" Rachel said to the caller. Her usual patience had been stretched thin, between trying to keep Walter from falling to pieces and dealing with the endless stream of callers. It went unspoken between them that Rachel was finally pregnant, too, due only a month after Kylie would have given birth.

"I'll let him know you need to speak with him. Now stop calling." Click.

Walter got up to head for the shower. The minister had reminded him how important it is to not only get up each morning, but to resume his routines as much as possible. If it had been up to Walter, in fact, he'd head back to the office today. But Rachel had given him such a look when he'd mentioned the possibility last night, he'd decided against it.

Walter had just finished dressing when the doorbell sounded, and, as he stepped into the hallway outside the master suite, he heard his sister opening the front door, and then a voice. Male, with a nasality that suggested a cold.

"You are totally inappropriate," Rachel steamed.

Walter crossed the kitchen into the formal dining room trying to place the familiar voice. As he entered the front hall, he saw Rachel doing her damnedest to push poor Randy Hall back into the snow that had blown onto the front porch during the night.

"Walter!" Randy called over Rachel's shoulder. Rachel was taller than him. It appeared she was also stronger.

"Randy? What the hell are you doing here?" Walter put a hand on his sister's shoulder. "It's okay, Rache," he said. "Let him in. The guy's freezing."

Rachel rolled her eyes but stepped past him and headed back to the kitchen, leaving the front door ajar and Randy on the threshold.

"Christ, Randy," Walter shook his head. "Come in out of the snow."

Randy mumbled a thank you and then an "I'm sorry" before following Walter to the dining room. Rachel came in with two cups of coffee. "I'll bring your breakfast in here when it's ready," she said. She was mad at him. Walter could tell. He thanked her anyway.

"So why did you come out here, Randy?" Walter asked him again.

Randy slid a glance toward the kitchen then leaned toward Walter across the table. "Veregro's killing everyone," he whispered, "one mouthful at a time."

Walter couldn't help it: he snickered. "Have you lost it, Randy? Do I need to call the white wagon?"

"I found your notes," Randy said.

"What notes?" Walter asked him.

"2012, when you worked in the lab."

Walter's coffee cup was halfway to his mouth, but he hadn't taken a sip. "You read them?"

"Not just that. I reran the tests."

"What for? Does Jim know?" Jim Baker, the R&D VP, was Randy's boss.

Randy didn't answer, instead pushed his glasses back up his nose. "I ran the tests, Walter. All of them. On rabbits. Mice. Hamsters. Rats."

"What a waste of time. If you read my notes why didn't you come talk to me first?" Walter stopped talking as Rachel came into the dining room, set a plate with eggs, bacon, and toast in front of each of them. She gave Walter a meaningful look. He responded with a smile and a nod instead of what she'd hoped for. She shook her head once, and left them.

Randy looked at the food as if nauseated by it. "Your results were conclusive. I've proved it."

"There were no real results. It was a pet project. There weren't the resources to validate the claims suggested. You'd have to run those tests cross-species, over three or four generations. How could you do that?" Walter picked up his fork and shoveled into his eggs. "You couldn't," he concluded, and took a bite.

Randy toyed with his fork moving his food around the plate. "Sure you could." He glanced at Walter, as if to make sure he was paying attention, then continued. "It's been my pet project, too. Did it on my own time. That's why Baker doesn't know about it. I ran your tests on four generations of each species, over the course of several years. The bottom line is that everything you edged up to but didn't solidify, all of it was conclusive. I thought you would be happy to hear this. I picked up where you left off."

"Look, I don't really remember all those tests," Walter said. He took hold of his index finger, then bent it back. "Number one, I left the lab." He bent another finger. "Number two, I got promoted." Another.

"Number three, I got married. Listen, Randy. It was a hell of year. I forgot all about those tests." He released the bent fingers all at once, wondering if Randy could tell he was lying. He'd never forgotten those tests. They'd kept him up, more nights than one.

"Okay. I get that. But now we have enough evidence that certainly compels us to do a full trial. Please Dr. Conroy, please just read my report. You're still a man of science."

Doctor Conroy. Man of science. It had been a long time since Walter had heard or even thought about those two truths. Randy was really working all the angles.

Randy continued. "Everybody wants to find a cure for Idiopathic Infertility Syndrome. If Veregro finds out its products are the cause, they'll want to fix it. They can't save and kill the world with their products at the same time."

Walter held out a hand. "Whoa there a moment, Randy. Who said anything about Veregro's products causing infertility?"

"It's fully covered in my report, the results in my testing are gender-dependent and—" Randy paused, and swallowed, "—conclusive."

Walter turned his fork over, as if the answer were written on its reverse side. He studied it for a moment, then set it down again. "They're not going to test," he said.

Randy was startled. "How can you know that?" he asked.

"I'm on the board. I know how things work. Results from an unsanctioned pet project, green-lighting a major study? Not happening. You are way out of protocol. And it would take years even if the board bought into it."

"Even if the preliminary results are conclusive?" Randy said.

Walter shrugged. "Rodents and humans, apples and oranges. One dead rat does not equal a hospital super germ or MODS or IIS."

Randy leaned forward. "Women losing babies is hardly a hospital infection! When I did the autopsy, the fetus had shriveled to nothing. It dried up inside, like there was nothing for it to hang on to. What if that's not just MODS? What if it's connected to IIS?"

Walter looked off toward the kitchen to see if Rachel had heard, then cautioned Randy. "Keep it down. My sister's pregnant," he said.

"Well, what about Errol Foster?" Randy asked.

"What about him?"

"His campaign to find a cure for IIS, that's what about him. We can hand it to him on a silver platter."

"You're after the reward. That's what this is all about, eh Randy? Errol Foster's cash reward so his wife can provide him an heir?"

It was Randy's turn to hold out a hand in protest. "No, I hadn't even thought of that—"

"Of course you thought of it. Who wouldn't? It's a lot of money. Enough for a smart kid like you to leave Veregro and set up his own lab somewhere, if that's what he wanted to do. Maybe I should claim the prize."

"This isn't about me," Randy said. "It's about what killed your wife."

Walter was silent a moment. Not just quiet. Completely still, as if his whole body had stopped. "I know what killed my wife. I was there."

"And now we know why," Randy hurried on. "IIS and MODS, they stem from the same problem. Knowing why it happens certainly puts us closer to stopping it."

Walter continued as if he hadn't heard him. "Kylie went in for her regular appointment at the clinic, and broke out in a rash that night. It was the hospital sickness. That's what killed my wife."

"It starts before that," Randy insisted. "It's not the hospitals. It started in her genes, before she was even born. But maybe we can stop it from triggering."

"You know what? You're fucking nuts. If you had seen her...." Walter stopped. He didn't want to see it again, himself.

Randy pushed his glasses up his nose again, leaned across his plate with its untouched food growing cold. "If it's a hospital super-germ, why can't they contain it?"

Walter looked at him with no answer. Randy swallowed, hard, then pushed on. "Walter, it's us. It's Veregro. It's all a result of manufactured seed, sprayed with Veresate. The combination of our products is the cause of all of it."

Walter slapped both hands on the table, then leaned back and shook his head, decisively. "You've got it wrong, Randy. Veregro is feeding people. We're saving the planet from dwindling resources."

Randy nodded, then stood up. "Well, I'm sorry I barged in on you like this. Your sister said it was inappropriate for me to come here. I suppose she's right. I thought I was bringing good news." Randy paused, waiting for a response, but Walter was silent, so he turned to head for the door.

"Randy," Walter called from the dining room just as Randy reached the front door.

"Yes, Dr. Conroy?"

"Send me your data."

THREE

LOTTIE WINTERS HAD JUST STRETCHED OUT in the grass on the crest of one of the hills that overlooked the Orchard when she spotted someone down in the valley, moving toward the path that led up to her aerie. The person was still a moving speck in the distance, but definitely heading in her direction.

Lottie wasn't alarmed. While it was unlikely to be someone from Outside, the Orchies seemed immune from the waves of disease that lately seemed to plague the nearby Townies. IIP—Idiopathic Infertility Syndrome—which meant that even if women were able to conceive, they couldn't carry babies to term, was bad. But MODS—Multiple Organ Dysfunction Syndrome—was worse: It was killing people— gruesomely. And no one knew why.

Without getting up, Lottie tried to determine who might be coming toward her. There weren't many among the several dozen Orchies who knew where she went to get away from things. J.P. and Samarie, her nearly grown kids. Pete Selin, her right-hand man.

They also knew not to bother her. She'd wait until the person approaching got closer before standing to see who it was.

Meanwhile, Lottie surveyed her domain. The country surrounding the Orchard reached out in undulating foothills covered in tall grasses and clover, sloping into the oak- and fruit tree-dotted hills that hid this verdant valley from all but the most intrepid Outsiders. Until Lottie's father, Charles, had come along and made an offer its previous owner had snapped up, no one had found much use for this piece of paradise. Bad farming practices in the 20th and early 21st centuries had long ago depleted the soil, and the place was too far from San Francisco to make it a practical location for a commuter.

But her father had known what others hadn't— that, properly returned to nature, this secret valley would yield healthy foods year after reliable year, all with remarkably little effort. Now, thirty years later, the rehabilitated land spread out below, orchards dotting the hillsides, pastured animals wandering anywhere they chose. The twenty or so houses and outbuildings were hidden in copses of live oak that also marked several clear, cool natural springs. Some people might have bottled that water to make a fortune. The Orchies chose to keep it to themselves.

Lottie had first come to the valley as a girl because her father, fortune made and civilization abandoned, had pushed farther and farther west in this still-wild corner of the county, seeking a place both to be alone and to apply the teachings of nature he'd studied over the years. While her father had died years before, his legacy had proved his theories right. Below Lottie lay the many miles of the hidden valley, protected and kept apart from the outside world that, like her father, Lottie had now turned her back on.

When she'd left for undergrad work at Berkeley and then law school at UCLA, she'd had every intention of proving her father wrong. Lottie would be a success in the world he'd abandoned. Far too soon, though, she'd learned the outside world chewed you up and spit you out without even noticing you'd existed. Lottie had returned home to her father's valley with her brief taste of success fouled by both scandal and the bitter taste of failure.

But success in that world, she reminded herself now, was not true success. Just look at what she'd come home to. Birds and wildlife thrived here, and, above the hills that ever since the ocean had crept further inland as the polar icecaps receded, remained green year-round, hawks and osprey rode the thermals, occasionally swooping in on unsuspecting prey.

"Mom?"

Lottie had forgotten the approaching figure. Samarie stepped out from between a pair of scraggly hilltop oaks, a tall, self-possessed seventeen-year-old wearing a loose cotton dress and handmade sandals. Samarie's easy beauty—her wheat-brown hair swinging straight and long, her wise brown eyes framed by long lashes—always flooded Lottie's heart with gratitude. This child—both her children—were products of that other world. Thankfully, it had been this world that had molded them.

"Why aren't you in school, young lady?" Lottie asked. Samarie sat, then gave her mother a grin.

"Pete got me out of class. He asked me to find you. There's a guy here asking for Charlie Winters."

The name came like a sudden gust of cold wind. Lottie sat up and hugged her knees. "Hmm," Lottie uttered, looking at the ground.

Samarie sat down. "It's a little weird, huh? Someone asking for Grandpa, like he's still here."

Lottie grew angry. "Why hasn't Pete sent him away? What's he thinking?"

Samarie shook her head slowly. "He told the guy to get lost but he didn't go. He sat down and started making notes or something. So Pete asked me to find you. He thought you should know."

Both Lottie and Samarie looked toward the place in the valley where they knew the entrance to the Orchard was hidden beneath a stand of oaks. Lottie tried to conjure up every man from the outside world who might come looking for Charlie Winters. She glanced again to the ground, plucked a wild flower and twirled it between her fingers.

Samarie tickled her arm. "Earth to Mom," she teased.

Lottie turned to her. "Who is this guy? Pete asked his name, right?"

Samarie shook her head. "I didn't ask, Mom. Pete asked me to find you so here I am."

Lottie stood, then offered an arm to pull Samarie to a stand as well. "All right. Well, put your game face on and let's go scare this guy out of here, whoever he is," she said, forcing a smile.

Once Lottie emerged from the oaks into the high grasses that the cattle favored this time of year, Samarie came up next to her and they walked easily, side-by-side, weaving among cows and their calves. A distance off, the bull looked up, then resumed his grazing. Lottie and Samarie crossed the field with their hands lightly brushing the tops of the grasses.

Pete, Shooey, and the man—but he was hardly a man; Lottie pegged him as the same age as her son J.P., twenty—were sitting on the stone steps of the dining hall. The stranger looked uncomfortable between the two older men. Both were in their fifties now, Pete's strong features having hardened into a rugged kind of handsomeness, dark-skinned Shooey nearly always smiling now that his demons from the third Gulf War had faded into the past. Samarie waved to them, then moved off. Lottie went to the foot of the stairs and crossed her arms.

The kid stood—Lottie gave him extra credit for doing so—and held out a hand. "Simon Keller. I'm a reporter. Maybe you've seen me on the Channel 4 news?" Lottie responded with a blank stare. Simon reached into his coat pocket. "Here's my NPAA pass," he stuttered. "I also write a blog. Have you heard of SimonSays.com? I blog about old stories that people shouldn't have forgotten."

A warning signaled in Lottie's temple, the same place it always did. She shook the proffered hand but didn't offer her name. Instead, she crossed her arms again and shook her head at Pete. She didn't move to sit down, and Pete and Shooey didn't move to get up. "Such as?" she asked, her voice wary.

Simon reached behind him, pulled a plexi tablet from a backpack and touched its screen. Nothing happened. He touched it again. The screen remained a window on the hand that held it.

Shooey laughed. "Ain't gonna get no signal here. We're off the grid," he said. "You're just gonna hafta tell us your story 'stead of showing it."

Simon touched the face of the clear tablet a few more times, as if he might perform some alchemy that would bring it to life. But the screen remained empty, and finally, he turned and put the tablet into his pack again. "Angell Farm," he said.

Pete stood, unfolding his six-foot-something frame so quickly Simon took a little jump back, into Shooey's knees. Shooey stood, too, and took a firm hold of Simon's elbow.

"That's enough of you," he said, a trace of a laugh still in his voice, because that was how Shooey talked.

"But...but..." Simon sputtered.

Pete had already picked up Simon's backpack and now held it toward him, swinging it by a strap. The half-tucked-in tablet tilted out precariously, and Simon made a leap toward it just as Pete swung the pack out of his reach.

Cat and mouse, Lottie thought. But she didn't try to stop it, even when Pete cast a quick look in her direction. She shook her head so quickly only Pete noticed.

Pete swung the pack slowly. Simon strained toward it. Then, just as the tablet was about to fall out, Shooey reached out a big hand and snatched it. He held it awkwardly, by a corner. "Hard to believe people think the world's in here," he said. He held the tablet out toward Simon, who grabbed it and hugged it close. Shooey laughed.

Pete handed Simon the pack then took his shoulder. "You ready?" he asked.

Simon tried to shake off the hand, but he was smaller and thinner and it wasn't any use. Pete encouraged him down the wide stone steps, past Lottie. Then Simon stopped and turned back to her.

"Are you Charlie Winters?" he asked. He was squinting at her, as if he were trying to match her to someone he'd seen.

Lottie felt a chill, wrapped her arms tighter. "Not anymore," she answered. Samarie opened her mouth to ask her mom a question then quickly shut it. Lottie turned and walked away, before Simon could ask her anything else. As far as she was concerned, Charlie Winters had been buried a long time ago.

Walter crossed to the door that led down to the basement and switched on the light.

The damp smell that greeted him was both familiar and surprising, familiar because it carried him immediately back to the basement of his parents' house, surprising because he couldn't remember the last time he'd set foot in his own. Kylie went down there—it was where the washer and dryer were—but he hadn't ventured down the stairs to his workshop in years.

But by the time he'd crossed the hopefully named playroom, his hand moved automatically to the light switch by his workshop door. Cobwebs danced between the bankers' boxes he'd stacked on the shelves above the long workbench, and Walter took an equally cobwebbed broom from the corner and pushed them away.

The boxes were labeled, but only Walter knew what each one held. Veregro might purge the mainframe files, but Walter had hard copies of his own, right here in River City.

He set up the stepladder, then climbed up and retrieved the box labeled *Charlie's Angels*. He wondered, as he had in the past, if his private joke was too transparent. Maybe he'd change the label when he finished rereading the files inside.

Walter dusted the seat of his work-stool with a rag that stilled smelled faintly of turpentine, then sat down and removed the lid from the box. The files smelled musty, and some of the pages had yellowed, but it was all there. Walter began at the beginning. He wanted to be sure he didn't miss anything.

At first, he hadn't even wanted to touch the yellowed clippings, it had been so long since he'd seen—or smelled—musty newsprint. But curiosity about why he'd saved them in the first place had won out, and, one by one, Walter read the stories that had recorded his life eighteen years before.

FARMER KILLS WIFE, KIDS, SELF

Des Moines, March 4, 2012. Des Moines County authorities confirmed today that central Iowa farmer John Angell, 42, killed his wife, Christine, 40, and three of their four children: Katherine, 12; Louise, 9; and Mary, 5. Angell apparently drugged his family's iced teas at dinner, and all slept soundly as he carried out his grisly task with the rifle he'd brought back from his service in Afghanistan.

A fourth child survived, but authorities declined to give details. Calls to local hospitals were not returned.

VEREGRO FORCED LICENSE AGREEMENT ON IOWA FARMER

Des Moines, March 5, 2012. Henry Kallen, Des Moines County Sheriff, today confirmed that an unsigned license agreement from Veregro was found at the home of John Angell, the farmer who shot and killed his rural Iowa family and himself on March 3.

Farmers all over the country have been required to sign such agreements, which include annual licensing fees to use Veregro seed.

According to Brad Stark, the owner of the Boone City Merc, Angell had not purchased Veregro seed, preferring to use, Stark stated, "what his daddy did."

"They had their lawyer up here waving papers and their scientists waving beakers and the farmers weren't having none of it," Stark told reporters. "But in the end, they had to sign, 'cause that seed is patented, you understand? John Angell, he thought it was wrong.

"You don't ever expect a man to go quite so far for something he believes in," Stark went on. "Especially a man you've known your whole life. But if you're a farmer and they want to take over your farm? It's a hard call. And it's a damned shame."

VEREGRO LAWYER ACTED ALONE

Omaha, March 8, 2012. In a press conference this morning, Veregro CEO Bob Howard announced the firing of assistant corporate counsel Charlotte Winters after it became clear that the young lawyer had been forcing licensing agreements on farmers without the company's knowledge.

"We at Veregro are committed to our friends in the fields," Howard asserted. He was surrounded by Veregro board members, including majority stockholder Foster Media founder and CEO Errol Foster. "Charlotte Winters' enthusiasm got in the way of her better judgment. Her employment with Veregro has been terminated.

"The death of the Angell family is a tragedy which we at Veregro mourn along with the rest of the country. The best way we can remember them is to make sure the same mistakes aren't made again."

Howard then introduced Walter Conroy, Veregro's new vice president of communications. "We wanted someone to oversee our communications who understands both our product and our commitment to conquering world hunger," Howard said. "Walter comes to us from the trenches in our testing labs. He's not just the best man for the job, he's the only man."

In prepared remarks, Conroy reiterated the company's commitment to learning from their former employee's mistake. "Veregro won't forget the Angell family.

That's why we've named our new global seed initiative the Angell Family Food Fund," he announced.

Eighteen years was a long time. Even as he read reports he knew he'd authored, it was hard for Walter to connect the man he was today with the one he'd been back when the Angell Farm debacle had happened.

Walter had been assigned to test the crop at Angell Farm because Veregro required a license to grow its patented seeds. Naturally, some farmers grew the seed without a license because of simple cross pollination: Winds carried pollen from one plot of land to another, as it had since time immemorial. But now, nature carried a Veregro patent on its breeze. Infringement meant farmers' fields could be seized by Veregro if their owners refused to destroy the patented crops. Of course, they were welcome to license their newly contaminated product. Just sign here.

None of the farms Veregro set its sights on stood a chance. It was merely a matter of time and due diligence before a farm was found to be using Veregro's patented genetic seed product, even if the farmer hadn't planted it. Solidly backed by testing—that's where Walter came in—it was a done deal.

To complete their *fait accompli*, Veregro assigned a young hotshot lawyer to follow Walter's test results with some legal findings of her own. Charlie Winters, twenty-four and newly anointed by the California State Bar, swept into Veregro with the fervor of a missionary: ambitious, pro-ag, pro-feed-the-world. Why? Because she had seen how the farming on her father's land changed lives. Now she wanted to feed the hungry in the ways she had heard Veregro did. Its campaigns were earth-changing, and Charlie wanted to be a part of them.

Charlie Winters had a whole dossier of farms to wrestle into a stranglehold until they cried uncle, but she had to start somewhere. She decided her first project would be the Angell Farm, where Walter's team had already taken samples.

Charlie's passion about Veregro amused Walter, but he didn't share it. He'd been at the lab nearly five years by then, and he liked his work—he truly believed that science was the foundation of everything. But he was less idealistic than Charlie. Nothing could be that good. Nothing manmade, that is.

By the time Charlie appeared, Walter had begun to notice that, at Veregro, business and science intersected. Definitive results were buried, or shuttered behind mumbo-jumbo. Once, as he read a final report on his own testing, he saw that the results hadn't been reported in their entirety, but rather to suit the spin of the report. On top of that, the FDA rubber-stamped every test Veregro sent their way. What if they were wrong?

Charlie once called him cynical. "A cynic's just a jaded idealist," he retorted, sorry as soon as he'd said it. The look on her face made him wish he could take it back. It was a hell of a thing to say to an idealist. He should have let her learn the way he had. Well, she did, in the end. She'd learned it better than he'd ever intended.

Meanwhile, Walter had begun his own quiet tests on the side. He didn't quit: Veregro wasn't tainting the drinking water with toxins, or chugging black smoke into the sky. In fact, as Charlie loved to remind him, they were increasing the food supply. They were feeding the hungry.

And yet, as he tested generation after generation, the results suggested Veregro's motives were perhaps more mercenary than charitable. Walter had been about to talk to his department head when the Angell Farm shit hit the fan. Shit and blood.

FOUR

ERROL FOSTER TAXIED HIS JET onto the Santa Barbara tarmac and radioed his readiness to the tower. He loved flying the Eclipse 500 he'd picked up for a song. He flew it to the cities where the various arms of Foster Media were headquartered. He flew it to the cities where the many boards on which he sat met. He'd even flown it to Peru, to pick up Sylvia Chavez, the surrogate he'd handpicked to carry his and Alice's child to term. Sylvia had been terrified. Errol had laughed.

"Two-Alpha-Niner, you're cleared," the tower told him. "Give my regards to Nebraska," the voice added with a sarcastic laugh.

Errol ignored the jibe. Veregro was headquartered in Omaha. As majority stockholder, Errol not only never missed a board meeting, he flew in a few days early and got the scoop before the meeting even began. It was his money, his rep. And, as with everything he did, Errol preferred to get his information firsthand. There was only one person Errol trusted: himself.

Errol taxied onto the runway, aimed, and pushed the throttle forward. The jet responded like a woman, a young and pretty one, purring its contentment as it moved faster and faster down the runway, and then, at a light touch from Errol, rose into the air. The ocean-invaded marsh beneath him grew smaller, and then he nosed through the marine layer and it was just him, his jet, and the sky.

Once he got to cruise altitude, Errol set the autopilot, then pulled his tablet from the briefcase on the empty seat next to him. In two swipes, he'd arrived at his latest Public Service Announcement, offering a cash reward to anyone who uncovered the cause of IIS, the mysterious infertility that was sweeping the country. Alice, his current wife, had miscarried three times before she'd told him no more. After a great deal of research, Errol had found Sylvia in Peru, a country as yet unaffected by the sickness. He'd have preferred to father a child with his beautiful Alice the old-fashioned way. But Sylvia was young and healthy, and had already carried their embryo far longer than Alice had those old-fashioned babies.

On the clear screen of Errol's tablet, a hologram of Errol appeared, its mouth moving but without sound. The jet might be quiet, but Errol still had to Bluetooth to his earphones to hear himself. On the video, he was sitting behind a wide desk that was part of one of Maya King's sets. Actually, they were his sets, since he owned the studio, the network, the grid, and everything associated with them. Whatever: It was a good looking set. That was why he'd picked it.

"I'm Errol Foster," he said on the video, hands folded easily on the desk, eyes focusing straight into the camera. He was wearing a dark blue turtleneck that set off his eyes, a camel blazer. He looked pretty damned good for a guy in his late forties, if he did say so himself.

"My wife, Alice, and I have been trying to conceive for years." The screen cut away to a photo of Alice, taken back when she was still modeling for Lauren. Back before all you could read in her eyes was pain and disappointment. "Sadly for both of us, none of our babies could be carried to term."

With the camera following, Errol got up from the desk and came around the front, settled casually against it. "I know a lot of you have suffered the same kind of repeated loss. That's why I'm offering a $500,000 reward to the first person to come up with a cure for our nation's growing infertility. We're all in this together."

Now petite, blonde Maya King appeared from Errol's right. Errol hadn't been expecting her, but, as always, her appearance lit up the video. Maya's talk shows garnered tens of millions of viewers, every day. If Maya blessed a person, a product, or an idea, it was as good as gold.

It was Maya who'd alerted Errol to the property next to hers in Montecito and, after he and Alice had renovated and moved in, she and Alice had made one of those quick and deep connections women seemed to specialize in, and which Errol couldn't understand.

On the video, after giving Errol a quick hug, Maya turned to the camera. "I'm here because Alice Foster is one of my best friends," she said. "That's why I'll meet Errol's 500 and make it a million. For Alice. For all of us. Let's find a cure."

The camera cut to Errol, who was smiling like the village idiot. Errol paused the video, checked the jet's instruments, then keyed in the video director's tablet on his screen. The director answered from a set. Errol could see all the activity around him. "Hey, Errol. What's shaking?"

"You need to edit out that cut to me at the end of the new PSA for finding a cure," Errol said without preamble. "End it with Maya."

The director was used to Errol. "You got it," he said. "You want to see it one more time before we broadcast?"

"What do you think?" Errol growled. He disconnected from the director without waiting for a reply. His altimeter reading seemed to be fluctuating. He tapped the glass, and the dial steadied.

Errol patted the dash. Sweet old jet. Of course she was getting a little touchy. Who used real dials anymore? But Errol preferred them to digital readouts. Older jets, younger women: a formula for happiness. That, and a shitload of money.

The flight path indicator said he was over Vegas. Errol leaned left to look out the window at the sprawling desert expanse whose daytime forlornness disappeared as soon as the sun went down. It reminded him of Alice, whose moods lately fluctuated so much he wondered if someone had snuck in a stand-in for the sweet girl he'd married.

He was sure it was about the baby thing. But even if a cure wasn't found, Sylvia would carry their baby to term, and Alice would be herself again. Once they'd determined Sylvia's insemination was successful, in fact, Alice had insisted on setting Sylvia up in one of their guesthouses. Now she spent most of her time with the young Peruvian, and Errol hardly saw either one of them. Well, that was fine with him, so long as they got their baby out of the deal. Sylvia would get her money, and then he'd fly her back to Cusco and get her out of his life.

Sylvia. For a brief moment, Errol closed his eyes to recall her lush curves and dusky skin. Then his eyes flew open. Sylvia was a wild card, and Errol liked to be in control. He ran his fingers lightly over the throttle, and the jet purred. Errol smiled. Yeah. Control.

Samarie and Rafi lay together in each other's arms on the sleeping bag they'd laid out in the abandoned bird blind's back cubby. She imagined that it used to be a pristine shelter where people would gather to observe the wonders of Mother Nature. But now it was just a dirty place, left to the whim of the weather and the neglect of mankind. It wasn't the most romantic spot to lose her virginity, Samarie thought. But where would be better? Maybe under a quilt in a room at the Orchard, on some drizzly day when her mom was doing Grounds rounds? Maybe in a windowed hotel room overlooking the ocean, stars flooding the room, setting her to tingle? That was how she'd imagined it, over and over again. But that was the past.

Three weeks ago they made love for the first time in this spot and transformed it into a holy place. Since their first night together, Samarie stopped dreaming of fancy hotels; instead she longed to return and relive that first experience. Rafi's gangly embrace made her feel somehow older and wiser. They became closer and closer each time they met in their spot. The sounds of their friends laughing and talking on the other side of the wall were a million miles away. The solitude was intoxicating.

The kids tried to meet up once a week, gathering in the large covered building. Birders hadn't ventured this far west since the tsunamis in the '20s, but the concrete blind had survived the rogue waves, and now provided shelter for the secret gatherings. Maybe the Townies' parents wouldn't have minded, but the Orchies suspected theirs would, in their mild hippie way.

Tonight, as on other nights, the kids brought sleeping bags and decks of cards, booze Townies got from their parents' liquor cabinets, and bags of beef jerky the Orchies brought from home. One of the girls had stashed a domino set beneath one of the shelves that jutted out from the windowless openings, and while a few of them played their hands, Bren interpretive danced her way through it. The bright yellow truck the Orchies had "borrowed" was backed up to the open bay, and music played out its speakers. A joint glowed its way around the circle.

A planet—or a universe—apart, Rafi and Samarie lay in the moonlight that poured through the cubby's windowless cutout. The way Rafi's lips touched Samarie's neck made her wonder why people did anything else, ever. She could be one of those stars, twinkling like they did on a clear autumn night. She pulled Rafi's face closer and pressed her body up to his. Nothing else in the world existed. No one mattered. Time stopped its ticking. Everything was blue. She said yes to it all.

When they finished, and lay in each other's arms, Samarie laughed. "This is illegal in some states, you know."

"Yeah, in the Bible Belt, maybe." Rafi kissed her nose and smiled.

"I'm serious. Can you imagine waiting till you're married?"

"I've waited long enough for you," he said. Samarie nuzzled his collarbone.

By the time Rafi and Samarie finally peeled themselves out of their sleeping bag and joined the others, the volume of the music had been lowered, and the fire outside was glowing. The group sat around it, talking quietly. A car, which they recognized as their friend Chrissy's, drove up. Chrissy got out of the car, alone.

"Hey Chris," one of the girls greeted her. "Where's Jordan?"

When Chrissy came into the fire-lit circle, Samarie saw that her usual smile was missing. "Jordan's not coming," Chrissy said. "His mom. She...died. Today."

A chorus of anguish and surprise rose up from the group. "What happened?" several voices asked at once.

"She got it," Chrissy said.

Samarie tried to imagine Jordan's nice mom, dead. But she couldn't imagine anyone dead. The only person she knew who'd died was her grandpa, and that had been too long ago for her to remember.

But what was "it"? "Got what?" she asked.

Chrissy gave her that look the Townies sometimes gave the Orchies, the *Don't you people know this is the 21st century?* look. "The sickness," she said. "MODS.

It's all over the country. But this is the first time someone's died of it here. Someone we know."

Todd spoke up. "It's totally fucking scary, is what it is."

"What are you talking about?" Willy asked. His mother, Amanda, was one of Samarie's mother's best friends at the Orchard. "I don't get it. A sickness that's killing people?"

Todd shook his head in seeming disgust. "Jesus, Willy. Don't you know? It's all over the news. People can't keep food down. Or they break out in weird rashes, or worse. Women keep losing their babies. People are dying. Like Jordan's mom."

Samarie was visibly startled, and Rafi put his arm around her shoulder. "That's so horrible," she said. "Why isn't anyone doing anything? Why didn't we know?"

"Nobody's been sick at the Orchard?" Todd asked.

Samarie shook her head. "You know we don't get sick much."

"Well, is anyone pregnant?" Chrissy asked, a challenge in her voice.

Samarie thought about it, then nodded. "Three women. No, four."

Chrissy huffed. "Well, they'd better be careful. It's hitting pregnant women really hard. They've even got a different name for that."

Todd kicked some dirt toward the fire, then turned from the group and walked away.

"Is everything all right with Todd?" Samarie half-whispered to Rafi. "He's not his usual self."

Rafi looked down at a rock and nudged it around in the dirt with his sneaker. "His mom is sick," Rafi said as he looked up at Samarie.

"Oh no. Sick, like really sick, or like, gonna-be-okay sick?" Samarie asked.

"Really sick. And Todd's really close with her, too, you know?"

"I know. I can't believe it. Jordan's mom, now Todd's mom," Samarie said. "This disease is really out of control, isn't it?"

"Todd said it looks like she's aged ten years in the last week."

Samarie shook her head when suddenly there was a shout. She thought she heard her name.

"Shh." Samarie and Rafi both listened. "It's Bren," Samarie said, looking around.

"Samarie, Rafi, you gotta come check this out," they heard Bren shouting.

They jumped up, Rafi grabbed his jacket from the ground and they ran to see what Bren was shouting about. Chrissy, Willie and Bren were huddled around Chrissy's tablet.

"Samarie," Bren said, waving her over. "Check this out."

"What is it?" Samarie asked.

Chrissy pointed her tablet toward Samarie. "It's a story about your mom, and the Orchard. It's on the news."

FIVE

ALICE AND SYLVIA WERE LYING side-by-side next to the pool when Consuela showed Maya in. Alice leapt up to greet her, and asked Consuela to align another lawn chair.

Maya, in diaphanous layers of royal blue silk, laughed. "I'm not dressed for pool sitting," she said. "I've got a show to tape. I just came to tell you before you hear it from somewhere else that I matched Errol's reward offer with one of my own."

Sylvia laughed, a delightful peal Alice had already grown fond of. "Oh, it is like the Wild West, with the good guys and the bad guys! I am liking it more and more!"

She was such an innocent. At thirty-five, Alice felt ancient in the face of Sylvia's relative youth, but it made her smile, too. "You're not in Cusco anymore," she reminded her.

Maya smiled at her allusion, then settled briefly on the side of Alice's lounge chair and put a hand on Alice's arm. "You're okay with me doing it? That's what I really wanted to ask. It was such a spur of the moment thing. Errol was using one of my sets—" she paused and laughed, "well, actually, one of *his* sets—to tape. I overheard him and just sort of bounded in. Next thing I knew, I was doubling his money."

"It will not hurt Mr. Errol to come down from his horse's top," Sylvia said. The other two women looked at her, unsure of what she was saying, and Sylvia tried again. "Is American expression, to come down from your horse's top. I hear it in a film. Or no. Your top horse. No, that is not right either. I like what it says, this expression. It reminds me of Mr. Errol when first I hear it. But now I cannot remember the words." She muttered in Spanish for a moment. "*Caballo. Su caballo... alto...*high! Your horse who is high!"

Alice and Maya burst out laughing. Then Maya stood and embraced Sylvia warmly. "Honey, not only are you right, but you're a breath of fresh air around here. It's about time someone brought *Mr. Errol* down to size."

Sylvia mulled this expression for a moment, then shrugged her shoulders. "Mr. Errol is of small size," she said. "But this means little. Because he is our *patron*. My *patron*," she corrected herself.

Maya stood and brushed the front of her skirt. "No. You had it right the first time, honey. *Our* patron. She turned to Alice. "Our little woman here's got lots to teach us," she told her. "Be sure to let me know if I miss anything."

After Maya left, Alice and Sylvia sat in companionable silence for a few moments. Then Sylvia turned toward Alice.

"Alice? Can I ask you the something?"

Alice rolled to face her. "Of course."

"Do you love Mr. Errol?"

Alice rolled onto her back again and slipped her sunglasses over her eyes. "Oh, Sylvia," she said. "That's a hard question.

Lottie was still fuming about Simon Keller the next morning. Pete tried to slip past her on his way to load the Orchard truck to make his town rounds, but Lottie had been watching from her porch and jumped up as soon as she spotted him.

"Do you have any idea what would happen if the Outside got wind of the Orchard?" she asked as she reached him.

Pete didn't slow, and Lottie trotted to keep up with his long strides. "I do," he said.

As usual, Pete's taciturnity staunched her anger. He knew it wasn't really him Lottie was angry with. Her father used to say she'd been born angry. "That's why you keep trying to save the world," he'd tell her. "But you can't do it by yourself. You'll see."

Pete strode off without her, and Lottie turned to walk back down the way she'd come. Set among the fruit trees were small homes with garden plots and trees, outbuildings and work sheds. Neil Thompson stood on his porch, surveying the day. He saluted Lottie as she walked by, while Bren, his teenage daughter, waved from the window, where she was trimming herbs.

Lottie waved back. Like her, Neil was second generation; his mother, Mary, had been one of the vets who'd found their way to the Orchard after Lottie's father had opened its gates to them.

Beyond the Thompson's place, the land swooped down and then gently up again to meet the hills that surrounded it. Orchards grew in spans up the hillsides, not in straight rows but rather following the curves of the hills and their turns of earth. Her father had seen it all before it had existed, and now, here it was.

During the third Gulf War, Charles Winters had fallen into a black depression. Reading newsfeeds moved him to tears; watching videos was a personal affront. While he continued to drive to work at Stanford every day, he was going through the motions. The world he'd always hoped would heal itself seemed lost.

Then, one Saturday afternoon, he happened to look out the window of his office in the Palo Alto Craftsman he'd bought with his first entrepreneurial windfall, to see his wife, Laura, trimming roses. Roses are beautiful, he thought. But why not grow more practical plants?

Charles turned on his computer for the first time in months, searched for and found information about organic gardening. Then he talked to Laura about digging up the small patch of front lawn and replacing it with leafy vegetables. Delighted that Charles was taking an interest in something besides the sorry state of the planet, Laura readily agreed.

After planting the tiny plot in the front, he moved on to the slightly larger back yard. Then he seeded both narrow side yards with native grasses, and began growing window planters of herbs.

The world may have disappointed him yet again, but Charles Winters couldn't help but come up with ideas for its betterment. His second windfall arrived when one of the biggest gaming companies in the world offered to buy the virtual reality game he'd designed. Charles' game had no warriors, no weapons, and no territory. Instead, it offered users a way to create their own world, so long as no one— and nothing—was hurt in the making. It had become a bestseller from its first day.

Not long after the buyout, Charles came home with the deed to the Orchard. Laura knew he'd been driving farther and farther north in his quest for unspoiled land. She just hadn't believed any more existed when he discovered the remote valley hidden between fog-wrapped hillsides.

After the tsunamis, the towns that still stood when the water receded were abandoned. The Tsunami Evacuation Route signs were long gone, as was most of the blacktop on the roads they'd marked. Only in the past few years had a few rugged individualists ventured back into what was left of the towns to homestead. They'd repaired the electric tension lines, dug new wells, and repaired the roofs and walls of the houses they chose to live in.

Charles had come upon one such town on a rare clear midsummer afternoon. He'd pulled up in front of the Mercantile, and, by the time he got out of his Jeep, half the town had come out to greet him. He asked if anyone knew of any land for sale nearby. A lot of land, he added.

Someone pointed. Another man hopped into the passenger seat of the Jeep and said he'd show him. He directed Charles east a few miles, back the way he'd come along what had once been the highway. Then he indicated Charles should turn left. They followed a grass track until it gave out, then got out of the Jeep and hiked over a rise. The man raised his arm as if in benediction.

The valley lay beneath them. Native grasses. Native oaks. Small coastal deer. And above, a pair of eagles, coasting an invisible thermal.

Ten thousand acres. Charles Winters bought every last one from an owner who was certain he'd rooked the city slicker.

A month later, as Charles eased the Jeep through the water-filled pockets that masqueraded as a road, Laura sat beside him wondering what she'd let him get her into this time. Then she turned to look at seven-year-old Lottie, kneeling in the back seat so she could peer out the side window. Lottie's eyes were wide with wonder, and Laura relaxed. It would be an adventure, just as she'd told her daughter before they'd left Palo Alto.

Once she got the household in order, Laura began digging in. But then her multiple sclerosis came back. Charles had hoped the clean start would keep it in remission forever, but it was too late for that.

At first, it was just muscle spasms. But then, all too quickly, she could no longer control her limbs. Within six months, Laura was gone, and it was just Charles and Lottie. Between an eight-year-old and the land he meant to live in harmony with, Charles didn't have time for the depression that had plagued him in the past. Besides, by then, the vets started arriving.

Just before Laura's illness had taken a turn for the worse, Charles had invited Pete Selin, an old friend who'd served in the third Gulf War now suffering from

PTSD, to come live at the Orchard. Charles imagined the Orchard would offer Pete the opportunity to garden, plant, harvest—and heal. Once Pete was there, and even though he wasn't the kind to gossip, he did let his platoon mate Shooey know.

Shooey was more garrulous. Word got out, and the Orchard grew into a little community of war-stressed returnees, planting and healing. Trips into town grew fewer and fewer, as the vets and their husbands and wives learned to build their own small, efficient homes. Together, they built the dining hall where the community could come together, and the workhouses where the livestock were taken so quickly and humanely the other animals were unaware of the buildings' purpose.

Putting their hands to work removed the vets from the war that continued to play in their heads. Nurturing a tree from soil to fruit became like focusing on breathing, tracing the quiet path of air through lungs, expelling it, and inhaling again. Moment by moment, life continued. Atrocities abated, and nightmares were laid to rest.

Throughout Lottie's teens, the people who came to live at the Orchard were her family. While a few came and went, a core group of families stayed, grew, and formed the foundation of the community. Pete. Shooey. The Starks. The Thompsons. Even after Lottie left and then returned, they never questioned her. They were simply there for her. The Orchard was home.

After Lottie's father died, the others turned to her. Had there ever been any question she'd carry on what he'd begun? Maybe in Lottie's mind. But confronted with the faces of those she loved, Lottie knew the Orchard was where she belonged.

When her father started the Orchard, the only wisdom he'd imparted to the men and women who joined him was to follow the curve of the earth. Watch the way it moves, he'd said, and then nurture its tendencies rather than fight them. Let the ground rest. Let yourself rest. Land that grows what it's meant to takes care of itself.

As she climbed onto her own porch now, Lottie smiled, remembering her father's words. All around her, his dreams had come to fruition. Under his guidance, the grasses native to the hidden valley had prospered, interspersed with fruit trees and small plantings of self-seeding lettuces and herbs. The hillsides were a mix of native oak and fruit trees, with the occasional bramble of berries growing beneath, while the perennial clover and grasses were perpetually fertilized by the bison, cows, sheep, and goats that grazed them, watered by the winter rains and summer marine layer.

Both Samarie and J.P. were doing schoolwork, and the house was quiet. Lottie sat in one of the rockers on the porch rather than go in. Before her lay her peaceful world, the world her father had made and that she'd inherited. Ordinarily, the mere sight of it could calm her natural tendency to worry. But today, something continued to nag at her. Was it the blogger, his invasion of her sacred space? Or was it something else?

Lottie considered a nearby pair of sheep, their lamb grazing peacefully nearby. The grazers were never milked, and when adult animals were periodically harvested for the Orchies to eat, they were moved far away from the other animals to deaths both quick and painless.

Like sheep to the slaughter. Where had that come from? Lottie tried to shake off the thought. Instead, it worried the stitch of anxiety she'd begun with and began to weave a fabric.

Something had changed. Lottie just wasn't sure what. She only knew that the gnaw of worry had turned sharper, and that she'd better pay attention.

When Pete pulled back in after making his town rounds, he stopped in front of Lottie's and motioned for her to join him in the warehouse. He was removing the last of the wooden crates from the back of the truck when she arrived.

Lottie grabbed a broom, then hopped up into the truck bed and began to sweep it out. It was the same truck the kids had driven into town the night before, although both Lottie and Pete were unaware of those clandestine get-togethers.

Lottie swept methodically, waiting for Pete to say why he'd asked her to follow him. When she could wait no longer, she stopped sweeping, and turned to him. "What's up?" she asked.

Pete was rubbing down the crates with an oiled cloth. "Things are bad in town," he told her, without looking up.

When he'd first arrived at the Orchard, quick movements had startled him. Then he'd taken on the job of being the Orchard's dealer, taking their goods to town, bringing the Townies' money back. Demand had grown, and he had a regular route now, some days going all the way down to Sausalito, others up the coast nearly to the county line.

Lottie put down the broom and sat down on the dropped tailgate. "What do you mean, bad?"

Pete kept polishing, but gave her a quick glance. "Every place I've been this week, they're talking about some kind of flu or something. They don't know what it is or how it spreads, but they're all scared shitless."

Lottie could feel the familiar chill beginning in her temple. "There've been flus before. We don't get them here. What's different about this one?"

"Today I heard a woman in Bolinas died. That's close, Lottie."

The chill became an icy stab. "Why haven't we heard about it before?"

Pete set down the box he'd been oiling and began on another. "They say it's all over the news." The Orchard didn't have a newsfeed. They didn't want one. Most news was manufactured for consumption anyway. It had little to do with reality.

"Did it just start?"

"It's been creeping along. A couple people hit. Old folks. Young kids. Ones with auto-immune disorders."

"That sounds like the usual pattern," Lottie said.

"It did to me, at first, yeah. But apparently it's not just the usual. Healthy people are going down, too. It happens fast."

"Do you know the symptoms?"

"I know that every stop I made this week, everyone I talked to had somebody sick. And a lot knew of someone who died."

"Died?"

Pete nodded. "Died."

Lottie shivered. "That's it," she said. "You're grounded."

Pete shrugged his shoulders. "Suits me."

Lottie watched him stack the boxes on the shelves, reconsidering. "But we don't really know what it is, do we? We don't even know if it's contagious or how it spreads. Our immune systems seem able to fight off just about anything the Outside tosses our way. Why not this, too?"

"How about we have a meeting?" Pete suggested, not looking her way. "You're the one always says we all have a say in decisions. I'm not sure I want to go back out there with this thing, though. It's bad, Lottie. I could tell by the looks on their faces."

Pete, scared? This was more serious than all the other scares. Lottie hopped down off the tailgate.

"Tell you what," she said. "I'll ask Samarie and Bren to help me spread the word. We'll have an all-Orchard dinner tonight and discuss what to do. And I'll talk to Felicia. She knows more about health than anyone here."

Lottie headed toward the big doorway, then stopped and turned. "Do you know the symptoms?" she asked.

Pete nodded, then looked away. "You don't want to know," he told her. "It's bad, Lottie. It's real bad."

Six

LATELY, ALICE HADN'T BEEN GOING into her office as much as she had back when she'd first started her cosmetics business, a few years before Errol had come along and swept her off her feet. She'd first avoided it after her second miscarriage, such a disappointment after the doctors had assured her the first was an anomaly. But work was what had finally pulled her out of that funk. Work, and Errol.

Alice's friends teased her about Errol. The man had a rep, not just as a Class A asshole, but as a latter-day Hugh Hefner who fancied himself a ladies' man when in fact he was just a dinosaur in Dolce & Gabbana. But Alice had always seen what others hadn't: Errol had a soft, sweet side. He never came home without flowers, and when he was away, which was often, she always arrived at work to find a spectacular arrangement on the conference table in her office.

But the third miscarriage—the one that came so close to term they'd already decorated the nursery—had changed things for both of them. Errol began to use the S word (for surrogate, before he found Sylvia and doubled its meaning), leaving Alice to feel as if she'd failed at the most important thing she'd ever attempted in her life.

Rather than come to bed, Errol would sit in his home office trolling the Cloud for countries where women weren't experiencing infertility or miscarriage, where babies were being carried to term just as they'd always been, where they were born healthy, limbs and organs intact. He'd come to bed so jazzed he'd awaken her with the news that Tongan women's birth rates had actually increased ("but so fat," he added, eliminating the possibility of a Tongan surrogate as quickly as he'd considered it), or the theory that the altitude in Nepal had made its women immune to whatever was causing the problems.

But Alice didn't want to hear it. She tried feigning sleep, but Errol shook her until she had to open her eyes. Finally, she told him outright that it hurt her when he talked like this.

The surprise on his face when she said that was so unexpected she immediately softened it. "I know you want a child, Errol. But you have to understand that I need time. I'm not ready anymore."

"But that's the beauty of it, Allie. You won't have to do a thing. You won't even have to lose your figure."

He ran a hand lovingly along her hip, and she flinched. He didn't get it. Or he didn't want to get it.

Once he settled on Peru, he flew there to, as he put it, scout. Scouting likely involved fucking all the prospects and seeing which one took. That was one thing Alice had never asked Sylvia. Maybe she would, one of these days. But not yet.

Because, to her surprise and, yes, delight, Alice liked Sylvia. She liked having the younger woman around, with her unusual combination of naiveté and directness. Sylvia brought a lightness to the house's many rooms that hadn't been there since early in the days of Alice and Errol's marriage, and her misunderstandings of English and America often made Alice laugh out loud.

But there was one thing about her life in Montecito that Sylvia had understood immediately: Errol. For someone who'd grown up in a convent, a hemisphere apart, she was surprisingly adept at reading beneath Errol's polished veneer. Just as Alice once had, Sylvia recognized the poor boy shuttered behind the mega-successful tycoon, the bloviating and bravado necessary to control his empire and everyone in it. Like Alice, she wasn't afraid to challenge him. Unlike Alice, she wasn't afraid of him at all.

Poor Errol. All he'd wanted was a vessel for his heir. Instead, he'd invited a hurricane into his home. It was only a matter of time before he began to see the waves.

Dinner at the Orchard wasn't always a group event, as it was at some communal farms. But that night, alerted by Samarie and Bren, every Orchie came to the dining hall. Each household brought a bowl or platter that they placed on the potluck table, which quickly grew to resemble a cornucopia of freshly harvested bounty.

Samarie and Bren sat at a long table with the other kids, Samarie's brother J.P. sliding in next to her despite her protests. While J.P. was three years older than Samarie, he behaved like a younger brother. A traumatic brain injury when he'd been very young had left him not only with a pronounced limp but developmentally disabled as well.

But J.P. possessed a sunny disposition that made him the light of any gathering. Now he snuggled against Samarie, his cheek sliding up and down her arm like a cat's. "But I love you, Sammy. You're the prettiest girl in the room."

"Stop it, J.P.," Samarie said, fixated on the door. She had gone looking for her mom before the meeting started, hoping to tell her about the news story, but couldn't find her.

Lottie showed up looking sweaty from a workout and addressed the crowd immediately as she made her way to her seat. "We have some business to discuss tonight, folks, so thanks for coming. I don't know about you all, but I'm starving, so let's enjoy our meal together first and then I'll raise the agenda. Like always, everyone who wants to be heard will be, so get comfortable." Lottie settled at a table with Pete, Shooey, the Thompsons, her best friend, Amanda Stark, and her husband the Rev, and Felicia Bruno, the Orchard's nurse-midwife. But as she began eating, her ears tuned in to the conversations that buzzed around her.

Miscarriages and fetal deformities...
Unexplained rashes, all over people's bodies...
Intestines turning on themselves...
And did you hear? A woman died in Bolinas...

Samarie pushed food around her plate with a fork, as if she might find her appetite hiding under some potatoes. She dropped her fork and went over to Lottie.

"Mom," Samarie said, interrupting Lottie's conversation with Pete.

"What is it, Sam?" Lottie asked.

"I need to talk to you for a minute."

"This really isn't the time, honey. I need to finish my conversation with Pete before the meeting. We'll talk afterward, okay?"

"Please. It's super important," Samarie pleaded.

"Later, Samarie. I've got a lot on my mind right now."

Samarie went back to her seat and waited for a lull in the conversation between Pete and her mom, but it never came. Lottie stood and tapped her spoon against her glass. "Time to discuss some business, everyone. I'm going to let Pete begin. Pete?"

Lottie knew Pete wasn't very comfortable speaking to the group, but she hoped he'd offer his firsthand knowledge. To her relief, Pete rose from where he sat next to her on the bench, and cleared his throat.

The group waited. No clattering of cutlery against plate. No whispers. Just anticipation—and dread.

Pete looked down at the table as he spoke, and the others strained forward to catch his words. "I don't know much for sure," he said. "Just that people are getting sick. Everywhere. I guess it's been around for a while, in the cities and stuff. But now it's here, in this county. Not just in the urban corridor, but in the towns I visit. All of them."

He paused, and the group waited for him to continue. Pete reached down and adjusted the position of his fork on his plate, aligned it to some internal precision.

Lottie reached over and touched his hand and he looked at her, then up at the others, waiting for him. "People are dying. Lots of people," he said, and a collective gasp passed through the room at the rumor, confirmed. "Healthy people, just, getting sick." He looked at Lottie, who encouraged him to continue with a low nod. "No one knows why. But we—Lot and me—we're thinking maybe we ought to quarantine ourselves for a while. See what's going on before we risk—whatever it is—ourselves."

"That's not fair!" Samarie cried. "It's not contagious." Bren patted her shoulder.

Lottie stood, and Pete quickly sat down again, relief written on his face. "We can't know that for sure, Samarie," she reminded her daughter. "It's better to be safe."

"My friend Jordan lost his mom but Jordan didn't get sick. And no one else in his family is sick," Samarie said. She looked to Bren. "Our friend Todd, his mother is sick. And no one else in his family has it! Tell them, Bren," Samarie pleaded, but Bren stayed silent. "The Orchard isn't a secret anymore. There's been a report on the news. My friends in town showed me."

The room reacted, voices filled the air.

"They know that no one gets sick here. That's what the story is about," Samarie continued. "We're all okay. Maybe we can help people!" The room went silent.

Samarie frowned, and Lottie tried to regain her composure. Any publicity was bad publicity. She found her train of thought again. "We've never quarantined before, not in all the years we've been here. For some of you," she nodded at those at her table, "it's been close to twenty-five years now. And I know how much some of us like our Town haircuts, the tools the Merc sells, but we've got to make sure that we're okay before we can help anyone else. And if they're reporting about us in the news, people are coming. Probably sick people. My vote is for caution."

Applause erupted from a few corners, showing Lottie their support.

"The way we live has meant that we don't usually need to sit down like this and..." Lottie paused to consider how they might come to a decision before she continued, "...vote. But now is the time to bring up any questions—even if we don't have answers— and then make a decision together. Majority wins."

She turned to Samarie. "I think you should start by telling us more about this news story that you saw." Samarie briefed the room and wisely left out any mention of Angell Farm. She knew it was best to tell her mom about that in private.

By the end of the meeting, there were only two dissenters against the quarantine: Samarie and Bren. Every other Orchie agreed that it was the best way to protect the Orchard. People could leave, but if they did, they wouldn't be able to return. Not until the quarantine was lifted.

As the meeting broke up, Lottie and Samarie looked at each other and Lottie motioned for the door. They set out on the short walk back home.

"You tried to tell me about that news story before the meeting," Lottie said, nodding. "I'm sorry I didn't listen to you. That was a mistake."

"I didn't mean to blurt it out mom. I wanted to tell you first."

"I know you did, Samarie. I'm not angry with you. You spoke your mind. That's a good thing. But I'm upset that the Orchard is on the news. I don't have a good feeling about it."

"There's something in that story I didn't mention in the meeting, Mom. It was written by that guy Simon who came here asking questions about that farm, Heavenly Farm."

"Angell Farm," Lottie corrected. They walked the rest of the way home in silence.

SEVEN

ERROL FOSTER DIDN'T LIKE to be kept waiting, and Marcia knew it. Ever since the Veregro CEO's secretary had ushered Errol into Bob Howard's inner sanctum, nearly an hour earlier, she'd been gliding back in bearing first coffee, then a bottle of Johnny Walker Green, and now, a fruit and cheese tray. Hand-hammered silver, Errol noted as she set it on the coffee table in front of him.

He'd already noted Marcia, the latest in a series of statuesque brunettes who would soon be bumped up the ladder when Bob grew tired of her. Marcia was built like a Barbie doll and had a voice like melted butter, but she was clearly smarter than the usual girl Bob preferred adorning his outer office. Errol figured she'd be gone by next week.

Marcia finished fussing with the vase on the tray and stood, offering Errol an expensive smile. "Mr. Howard just called to say he's on his way back from the lab," she said. "He should be here in ten minutes."

"The lab?" Errol couldn't hide his surprise. "Bob Howard's kept me waiting while he's playing at his lab?"

The smile dimmed an instant before Marcia relit it. "I don't think he was 'playing,' Mr. Foster. There was a call early this morning that one of the techs may have breeched protocol, and Mr. Howard went to see to it personally."

Errol slid forward on the couch, the better to pour himself a finger of Scotch. "What kind of breech?" he asked.

Marcia shrugged, beautifully. "I don't really know, Mr. Foster. But I'm sure Mr. Howard will fill you in as soon as he—"

The door opened and Bob Howard breezed in, all 6-foot-3 of him looking trim and fit in his custom-tailored grey pinstripe three-piece. "Errol! Sorry to keep you waiting. Marcia, thank you. You can go now."

Marcia slid out and Errol stood, offering a steely glare along with a handshake. While they'd met most of the previous day, and parted on a jovial note, Errol's anger at being kept waiting had soured that camaraderie. He sat down without their usual easy banter, then growled, "Tell me what's up."

Bob pulled one of the big leather chairs that faced his desk around to face Errol. "One of the lab techs read Walter Conroy's old test notes," he said. "Hey— pour me some of that, too, will you?"

Errol poured a second glass of Scotch and slid it across the coffee table. "Those notes were destroyed," he reminded Bob.

Bob took a sip of the Scotch and nodded his approval. "That's what I thought, too. But apparently there was a copy in the archives. The tech came across them during a routine file purge. They were encrypted but he broke the code—we're not sure why he was even interested—and read them. His lab partner thinks he's been conducting some tests of his own on the sly."

Errol gave Bob a cut-to-the-chase glare. "So what are we gonna do about it?"

"Already done."

Errol set his glass down. "Good. What about the rest of the lab? Are they gonna question it?"

Bob shook his head. "His lab partner says he was acting strangely, but that's pretty strange anyway. He was all broken up because one of his rats died. Apparently, he'd named her. No one is going to question it. He was a weird guy."

Errol moved his glass to the side and began to trace the wet circle it had left on the coffee table. "What else does the lab partner know?" he asked.

"Nothing, from what I can tell. When I questioned her, her knees were knocking so hard she sounded like Buddy Rich. But she's a talker. She couldn't keep a secret to save her sister."

"Tell me about the lab tech," Errol said.

Bob got up for his briefcase and extracted his tablet, then sat down again and swiped a few times to arrive at the file he wanted. "Randy Hall. Thirty-nine. Single. Born and raised in Omaha, but no family remaining. UNO Honors in zoopharmacology. We hired him before graduation and he's worked here ever since. Borderline Asperger's. Loner. Comes in early, works late."

Errol drummed his fingers on his knee. "You sure there's no one to miss him?" he asked.

Bob swiped the screen again, then shook his head. "Not even a cat," he answered.

"So it's done," Errol said.

"It's done," Bob echoed. He pushed his glass toward Errol. "Shall we drink to that?"

EIGHT

"MAYBE I CAN GO INTO TOWN with Pete." Baskets on their arms, Bren and Samarie were foraging for spring greens beneath the trees on an Orchard hillside.

"He'll never take you," said Bren. "Besides, he's only going one more time for supplies. He's not going to have room for a passenger."

"I can't just not go back. Rafi doesn't know why he hasn't heard from me. He'll think I'm mad, or that he did something wrong."

Bren shrugged her shoulders. "He'll figure it out."

Samarie shot Bren a look. "Do you not want me to see him?"

"No! I just don't want you to get in trouble. Or sick. What if we're wrong and the disease is contagious?"

"There was nothing wrong with our friends. You saw."

"Maybe it takes a long time. I don't know. I just don't want any of us to get sick, okay?"

"I can't believe you. People are fine. I thought you supported me."

"They're not fine, Sam. Jordan's mom died. Do you call that fine?"

Samarie set down her basket and grabbed her friend's arm. "What if something happens to Rafi? I have to see him."

"Does your mom even know about you guys?"

"No."

"Well, I don't think it's a good idea. Pete might talk to Rafi for you, but he's not going to let you go with him."

Samarie looked at the ground and wrung her hands. "I love him."

Bren rolled her eyes but set her own basket down. "Listen, Sam. Rafi'll understand. You can't be exposed to some weird disease. He'll get it. He'll know you're not avoiding him."

The girls picked up their baskets and resumed their foraging. "I know you love him," Bren said after a few minutes.

Samarie straightened to look at her. "You don't know everything, though."

Bren straightened, too, and narrowed her eyes. "What don't I know?"

Samarie looked around, as if the trees might have ears. Sometimes her brother J.P. followed them, but she didn't see him now.

Bren put her hands on her hips. "What? Tell me."

Samarie leaned toward her. "I might be pregnant," she whispered.

"No."

"Yes."

"It's not possible. You just—I mean, it hasn't even been—how do you know?"

"I should have started my period five days ago."

Bren laughed. "Five days? It's hardly time to start decorating the nursery."

Samarie frowned. "No. Really. I know. I'm Felicia's apprentice, remember? I've got all the early signs. Nausea. Emotional volatility...."

"Wishful thinking?"

Samarie grabbed Bren's hand. "Bren, look at me. You know me. I'm not a head-in-the-clouds kind of person, am I?"

Bren shook her head, slowly. "Is there some kind of...test? Some way to know for sure?"

Samarie nodded.

"And you already did it?"

Samarie nodded again. She'd snuck the test kit out of the clinic when Felicia wasn't there.

"Wow." Bren paused for it to sink in. "Does Rafi know?" she asked, finally.

It was Samarie's turn to shake her head. "What am I supposed to do? Send Pete to tell him?"

"I don't know. You always say you guys can read each other's minds."

"Jeez, Bren. That's when we're together. Seriously. How am I supposed to let Rafi know? He'll want to know, Bren. He'll be happy."

Bren set her basket on the soft ground and sank down next to it, patted for Samarie to join her. "Can you tell your mom?"

Samarie stroked the soft green moss at her side. "I kind of have to, don't I?"

"Jeez. She'll probably have a cow."

Suddenly, J.P. stepped out. He'd been behind a tree. How had they not seen him? Fear was written across his features. "Would Mom be okay, if she had a cow?" he asked.

Samarie grabbed his arm and yanked him toward her. "How long have you been here?" she asked. "What have you heard?"

J.P. tried to shake away from her. "That Mom's going to have a cow," he answered, his voice teary. "Let go. You're hurting me."

"What about...me? What did you hear?"

J.P. cocked his head, as he often did when he was trying to piece something out. "You love Rafi," he said, finally.

"And...?" Samarie prompted.

J.P. returned to his earlier fear. "What if Mom has a cow? It will hurt her!"

Samarie relaxed her hold and J.P. took a step back. "Mom's not going to have a cow," she assured him.

"But Bren said...I heard her. I've got to go tell her."

"J.P. No!" Samarie leapt up and reached for J.P.'s arm. But that was enough for J.P. to turn and start running toward the houses.

"J.P.!" Samarie called after him. "Come back! J.P.!"

"She's gonna have to find out sooner or later," Bren, still seated, said.

Samarie looked down at her. "Did you have to say that about her having a cow? You know how J.P. takes everything literally."

Bren stood and brushed herself off. "We didn't know he was there."

Samarie reached over and dusted a dried bit of oak leaf from Bren's shoulder. "We should have, though. He's always following us. It's my fault. I was so worried, I forgot."

"You'll need to tell your mom," Bren said.

Samarie nodded. "I will. But I think she'll be okay. I mean, she would have been my age when J.P. was born. And she wasn't married. We don't even know who our dads are. My baby will."

From a distance, they heard J.P. calling, "Mom! Mom!"

In spite of everything, Samarie smiled. "I guess I'd better head home," she said.

Bren returned the smile. "Let me know how it goes."

"Oh, I will," Samarie assured her. "Either that, or you'll hear the screaming all the way to your house."

Walter backed up Randy's data into a Cloud file before putting on a suit and tie and stuffing his tablet in his briefcase. By the time he arrived at the boardroom at 11:30, their lunch break was only a half-hour off.

Errol was the first to acknowledge his entrance. He stood and moved to the door, shook Walter's hand, and reached to grip his shoulder with his free hand. "I'm so sorry, Conroy," he said. "We're all sorry. How are you holding up?"

Walter nodded as he took his regular seat. "I'm okay," he answered. "I appreciate your concern." He set his briefcase on the table and extracted his tablet and the stack of papers Katrina had quickly printed when he'd surprised her by showing up at his own office. "Please go on. I have something to share with you, but I'll wait for the new business portion."

Errol and Bob Howard exchanged a look. "We're about there now," Bob assured him. "Are those printouts for distribution?"

Walter nodded and passed the stacks to Marcia, who stood and slid one first in front of him, and then before each of the six directors and the four VPs. Each acknowledged Marcia's proximity with a smile and a nod.

By the time she'd finished and sat, Bob had already flipped through his copy. Now he stood, pushing his chair back from under him so suddenly the others looked up. "Conroy," he said.

"Sir?"

The CEO turned to Marcia. "Sorry to double your work, honey, but can you collect what you just distributed?"

Marcia didn't question the order. She did as she was told.

Bob walked around the table until he stood behind Walter. He set his hands at the corners of the big leather chair.

"We've got a full plate today, Walt. Can hardly cover everything as it is. Let's discuss this in my office."

"It's serious, Bob. The board needs to hear it."

Marcia was waiting next to Errol's chair. His was the last presentation to collect and he flipped through its pages, scanning through parts, eyes settling on others. He didn't look up.

Walter continued. "We've got a chance to stop MODS."

The small smattering of sound there disappeared.

Bob had his hands on Walter's shoulders now. "We all know you've suffered a terrible loss, Walt."

Walter shook off Bob's hands. "It's Walter," he reminded him.

Now Errol stood as well, presentation in hand. "Walter. Let's you and me talk this out. Bob, you all can finish without me. Marcia, you'll come get me if there's a vote."

Marcia nodded. Errol walked over to Walter's seat and Bob stepped aside. "The man needs to be heard out, Bob."

"Of course," Bob said.

"You won't even miss me," Errol assured the rest of the group, and they offered a grateful murmur of laughter that eased the tension. Errol took the printouts from Marcia and led Walter to Bob's office, down the hall.

Errol went behind the desk and Walter took one of the leather visitor chairs. Over the credenza to his right hung a graphic from an old marketing campaign, the word "Teamwork" superimposed over a string of rainbow-colored genes that morphed from a DNA strand into a piece of corn.

Errol folded his hands and studied Walter's face. "You're a bold man to come to work at such a difficult time," he said. "You sure you're all right?"

"It's important," Walter said.

"I can see that." Errol sifted through the printout, stopping at a two-page spread of a flattened Earth. "Tell me. What's this?"

He turned the page so Walter could look at it right-side up.

Walter studied the visual Randy had sent. Each landmass was highlighted in either yellow or green. Only a few were green. Most of the rest were yellow, and a few very small countries had no color at all. Errol put a finger on Peru. It was green.

"The places that are green have either banned GMOs in seed, or they haven't yet bought into the technology," Walter answered. "The yellow countries have our product, either in Veregro form, or as developed by smaller companies licensing our patents. And they're spraying it with Veresate."

Errol turned the page back toward him. "What exactly are you saying here, Conroy?"

Walter reached down into his briefcase, set his tablet on the desk and opened it. He swiped a few times, then looked up at Errol. "Bottom line, the yellow countries have more miscarriages and higher mortality rates—"

"Of course they do. They're more developed. They have higher populations."

"Per capita."

Errol looked down at the graphic. "Go on."

"A long time ago, I ran some tests. Our product didn't meet FDA requirements, but it was approved anyway. I thought I must have missed something. I decided to run a few more tests, over several generations."

Errol's fingers drummed on the map. "And?"

"I never finished my tests. But the results were already becoming definitive. When a GM crop was sprayed with Veresate, abnormalities began to appear, several generations down the road. Our products were negatively affecting the subjects' immune and reproductive systems."

The drumming intensified. "Why did you stop?"

"You know the answer, Errol. I left the lab. I got promoted."

"And now? You think your eighteen-year-old tests are going to bring back your wife?"

Walter's surprise registered on his face for only an instant before he clicked off his tablet and stowed it back in his briefcase. "I guess we're done, aren't we, Errol?"

Errol stood and offered his best smile. "Why don't you leave your report here and I'll fill Bob in on your findings? Personally, I think it's a noble effort you're making in memory of your wife. But you're biting the hand that feeds you, taking aim at the company that's employed you for twenty years, and which, I might remind you, has become the world's breadbasket. You're making some serious claims here, Conroy. Do you understand what going public with this could mean?"

Walter finished stowing his tablet before standing himself. "I was right," he said. "I knew you'd be most interested, what with your personal campaign to cure infertility." He took a step toward the door.

Errol moved out from behind the desk. "What I'm interested in," he said, trying to keep his voice even, "is finding out what's behind IIS. I'm not looking for over-reaching junk science, patch-worked together in a moment of grief by some has-been lab tech to level at the world's largest charitable food source."

Walter walked to the door. Errol followed him, placed a hand on the knob.

"That said," Errol continued, "your report, spare as it is, is due serious discussion. I'll bring this to Bob, if not the full board before they all go home. Rest assured that we'll discuss the appropriate action. It may be that a trial is in order."

Walter waited for him to open the door but Errol didn't move. "I'm not the only one who's run these tests," he said.

Errol raised an eyebrow. "And who might your partner in crime—or should I say, glory—be?"

Walter pictured Randy Hall, pushing up his Band-aid-repaired glasses. "I'd rather not say," he said.

Errol met his eyes for a good ten seconds, and in that span of time, Walter knew he was right not to bring Randy into it. If they came after him, fine—he was a widower now. What did he have to lose? But Randy was still young. Idealistic. Well-meaning. This wolf would eat him alive.

The wolf was smiling. "Thank you for bringing this to our attention, Walter. Now, you take care of yourself, and let us take care of what happens next." Finally, he turned the knob, and held open the door.

As he crossed Marcia's office into the vestibule, Walter's hand slid into his pocket, where he'd slipped the sheet of paper with the encryption codes for the files he'd uploaded to the Cloud. He checked his briefcase for his tablet one more time. Then, rather than wait for the elevator, he headed for the stairwell. He couldn't wait to get a breath of clean air.

Pete stood at the gate, studying the variety of people who'd begun to gather there. He overheard a conversation between one young couple talking about the lengthy newsfeed about the Orchard and Simon Keller's blog. Many people asked Pete if he was immune to the disease but Pete said nothing. If he had it his way, he'd sell them some fruit and meat, but Lottie nixed the idea.

About ten feet away to either side of Pete, J.P. and Willy stood uncertainly. Lottie had dispatched them after Pete reported that the crowd was growing so quickly he might not be able to keep them from storming the gate. J.P. offered his lopsided grin when Pete glanced his way; Willy, stocky and muscular in his usual white T-shirt and jeans, crossed his arms and did his best to look menacing.

If anything, though, the crowd was unusually respectful of the boundary. The Goddies, as Pete had nicknamed one group in his head, spent most of their time creating ever-more referential signs. Their most recent cited the Ten Commandments, somewhat obscurely: *Thou shalt not place false gods before me.* Pete supposed they were referring to the conceit that man could alter DNA in order to produce genetically modified seed, a job they clearly felt should be left to their god.

The Latter-Day Hippies sported no signs, but were distinguishable by the periodic scents of pot and patchouli that blew Pete's way. Four generations removed from those who'd pioneered the movement, they'd adopted the parts they liked—free love, free pot, and (so Pete phrased it) free loafing—and ignored the rest.

Disgusted, he turned to yet another group, this one not yet coming together but rather comprised of solitary young couples like the one he'd interrogated earlier. Some of the women were clearly pregnant; Pete could understand why they might come to this remote edge of the continent to seek reassurance. Others were simply forlorn; Pete imagined they were the ones who'd miscarried, perhaps more than once.

And then there was the media. They were trying to blend in, but blending in was not something a talking head or blogger could pull off. For one thing, they either wore backpacks or messenger bags that held their ubiquitous tablets. For another, they were usually alone, scanning the others restlessly, ever seeking The Story. Male or female, they were young, hungry, and angry, without knowing precisely what they were angry about.

Pete remembered what he'd been angry about, all those years ago, before he'd shelved it for good. His country had used him, sent him to a backward desert halfway around the world and expected him to kill indiscriminately in the name of freedom and justice. And oil.

Initially happy to serve his country, Pete could pinpoint exactly when his disillusionment began. It was when he was assigned to the prison, and witnessed the indiscriminate torture of inmates who may or may not have committed crimes, women made to wear hoods and suffer verbal haranguings, men made to kneel naked while their nether-parts were teased with rifles. Pete shuddered even now, remembering it all. Fuck the world, he'd decided after his discharge. Who knows what he might have been capable of had Charles Winters not found him and insisted he come out here and help him start the Orchard?

More cars pulled up, parking haphazardly by the side of the gravel road that led to the gate. Newcomers emerged slowly, like butterflies easing out of their chrysalises, blinking in the sun, looking this way and that at the place they'd come to. Periodically, a movement from Willy, J.P., or himself would cause the crowd to turn their way, in unison, as if they were finally going to give them what they came for.

But what had they come for? Whatever it was, Pete Selin was hardly the person to deliver them from their affliction.

Walter waited until he was home to call Randy Hall, and when Randy didn't respond to Walter's zaps or answer his tablet, he tried his lab extension. He was surprised when a woman answered. Short, blonde, non-descript, mid-thirties. Behind her, Walter could see the stacked animal cages.

"Is Randy there?" he asked.

The woman hiccupped a response he didn't catch.

Walter fiddled with his volume control, although he suspected that wasn't the issue. "Could you speak more clearly?" he asked her.

The woman took a deep breath, then spoke in a rush. "I'm sorry. Who is this?"

"Walter Conroy. And you are?"

"Pauline Wozeck. I'm—I was—Randy's lab partner."

Walter's pulse raced. "Was? What do you mean?"

The woman whimpered. "Randy died. This morning."

Walter backed into a chair. "Died?"

"Of the disease," she whispered. "Of MODS." She sat down, too. Her chair creaked.

Walter leaned toward the screen, thumped the table with his palm. "That's not possible! I just saw him the other day. He—" He stopped himself from continuing. Randy had never said if his lab partner was aware of his tests.

"Well, it's true," the woman went on. Wozeck, she'd said. Pauline Wozeck. "Bob Howard just came down to tell us. He was here earlier, asking questions about Randy and what he was working on, but we never imagined why. I mean, Randy went home early yesterday, but I thought he was just sad about Daisy. Randy loved that rat like I never saw. I told him not to name them, but—"

Walter interrupted her. "Bob Howard came to tell you?"

"Wasn't that something? We all thought it said something about Veregro that Bob Howard himself would come down to break the news to us. We're all being tested now, to make sure we don't have it—the disease."

Walter was surprised. "There's a test for MODS?"

"I guess so. It's funny none of us knew about it, but we all were sampled." She pointed to her lab-coated elbow crease. "Jim Baker's running the results himself."

Bullshit, Walter thought. There was no test. Baker was just following Howard's orders. And Randy hadn't died of MODS. Walter knew far too well what the disease looked like. Randy hadn't had it. Randy had died of something else.

Just then, a message popped up in the corner of his screen. *Randy Hall*, it read. Walter shivered. A voice from the dead. He clicked it open. *IL*, it read. *JKL8743*. What the hell? There was an attachment, too. Big. Encrypted. Walter was pretty sure he knew what it was.

"Mr. Conroy? Did your mic cut off?"

"I'm here."

He needed time to think, but time, he suddenly realized, was probably something he didn't have much of. He'd mentioned Randy to Errol Foster. He'd passed Randy's test results on to both Errol and Bob Howard. That meant they knew Walter knew.

And if Randy was dead, it meant they'd stop at nothing to make sure what Walter knew didn't get out. Walter needed to get the hell out of Omaha. Now.

Pauline Wozeck was still watching him on his tablet. Did he dare ask her not to mention he'd called?

But of course, the phones were monitored. As a Veregro board member, Walter already knew that. The question was, was his personal tablet being monitored as well? Walter took a step away, as if the movement could distance him from the possibility.

"Pauline? Thanks. I'm so sorry to hear about Randy."

Pauline dabbed at her eyes with a tissue. "We're all so surprised," she said again. "But the fact that Mr. Howard came, and now you—well, that says something about Veregro, doesn't it? It says what a caring company it is—"

Walter interrupted. "Of course," he said. "Thanks again, Pauline. You take care now."

Pauline nodded. "Why, you, too, Mr. Conroy. You take care."

Walter severed the connection. If he was going to take care, he'd better get moving. Now. The question was, where? And how?

Walter turned on the private search function, then typed in Charlie Winters' name. Her full name,

Charlotte Winters. The first hit was the newsfeed he'd watched that morning. Walter opened his text program to make notes, then clicked to watch it again.

NINE

WITH CONSUELA'S HELP, Alice and Sylvia had turned the den into a meditation room. Maya's show on meditation had given them the idea, and they'd spent an afternoon wandering the house, considering each of the rooms in turn. They'd settled on the den because of its view through the trees to the ocean and the gentle wisteria-laden breeze that wafted in through the windows.

Alice's only hesitation had been that the den was supposed to be Errol's office. He never used it; if he needed to conduct a work-related conversation at home, he usually went out to the pool, or under the portal if it was raining. She'd been the one to select the cordovan leather furnishings, the rare antique books that filled the shelves. Errol would probably be relieved he'd no longer need to feel guilty for not using the space on which she'd lavished so much time—and money.

They enlisted Maya in their cause, then headed down to Santa Barbara in her Bentley the next morning. Shopping with Maya was a treat. Everyone knew her, of course, but then, everyone knew Errol, too. The difference was that everyone loved Maya, and Maya loved them right back. Maya was charisma personified. You couldn't help but get caught up in it.

At the home furnishings store where they stopped first, Sylvia was drawn to the brightest colored floor pillows—shocking pinks, luscious raspberries, vivid blues.

But when Maya said a meditation room should be furnished in soothing earth-tones, Sylvia demurred. They ended up selecting a variety of pillows in tones of taupe and gray before continuing on to the flooring section, where they opted for a renewable bamboo.

Next, they walked a few stores down to the Asian Lifestyle Emporium, where they chose several Buddhas, incense holders, and finally, the incense itself. Alice couldn't resist a tall antique brass gong, its resonance as soothing as a long, solo swim. Sylvia asked if they might purchase the equally resonant wind chimes, so tall and round they registered a deep bass. Maya, who'd wandered into a far room, returned bearing silky pajamas for each of them. "These alone will put you to sleep," she promised.

Now, Alice in her silver pajamas and Sylvia in her gold ones, they stepped into the new room for the first time. Sylvia picked up the felt-headed wand and touched it to the big brass gong, once. As it sounded, Alice felt a deep peace ease over her. She settled cross-legged onto one of the pillows, and Sylvia did the same.

Through the open window, Alice could see the tops of the eucalyptus across the canyon and the broad reach of ocean beyond them. The morning's fog was gone, leaving the Channel Islands visible beyond the long-abandoned oil platforms. Alice tried to stretch her mind across the ocean, all the way to the place where the Buddha had once lived. Then she remembered the syllable Maya had given her to focus on, and hummed it, softly. Next to her, she heard Sylvia chant her own syllable.

The next thing Alice knew, Errol was standing in front of her. "What the hell?" he was saying. His voice was quiet, but Alice could feel the anger in his words.

Sylvia was curled on her side, asleep. Her pajama top had hiked up enough for Alice to see that she'd begun to show, and she was at once filled with both longing and desire.

Errol pulled at the knees of his khakis then squatted down in front of her. "What is going on here, Alice? What has happened to my study?"

Alice could have responded as she usually did, defensive or apologetic. But something had changed while Errol'd been gone, and she instead smiled. "You never used it, sweetie. We decided to turn it into a meditation room. And see how well it's working?" She giggled. "It's put Sylvia right to sleep."

Errol stood, then walked to the incense and crushed its tip into the tray until it stopped burning. "It's probably that shit that knocked her out," he said. "And who knows if it's good for her? What the hell were you thinking, Alice?"

She wouldn't—or couldn't—rise to his bait. "It's a room for us to de-stress," she explained, nonetheless feeling anger gather in her gut. "Except now you're bringing all your negative man-energy in here."

At this, Errol finally laughed. "My negative man-energy? Christ, Alice. Who've you been talking to while I've been gone?" He crouched again and stroked her cheek. "You've always been pretty positive about my man-energy."

Alice flinched. She couldn't say why she had; but the reaction had been visceral. She hoped Errol hadn't noticed. When she put her hand over his, it was clear he hadn't. "Maya did a show about the value of meditation," she told him, keeping her voice low and calm. "Especially how good it is for pregnant women." She turned and kissed Errol's hand. "There've been studies, sweetie. And just look how relaxed Sylvia is."

They both turned to consider their Peruvian surrogate, her light snores punctuating the room's quiet. Then Errol offered Alice a hand to help her up. "Looks like we've got some time to be alone," he said, his smile gentle and beguiling.

Alice returned the smile, and hand in hand, they crossed the living room. Later, Errol thought, he'd have to have a talk with Sylvia. He couldn't have Alice poisoned by crazy ideas, or their baby compromised by foreign odors or customs. But for now, he led Alice to his bedroom. It had been a long time. Errol had missed her.

Lottie wasn't sure how long Samarie had been following her from room to room. But when she settled on the front porch with a cup of tea and a headful of worry, Samarie was only a few steps behind.

"You haven't followed me this close since you were four," Lottie said, patting the porch swing for Samarie to join her. "Everything okay?"

Samarie smiled and nodded. She curled into the free side of the swing and then Lottie gave them a gentle push that set them rocking.

"Any idea why your brother keeps talking to me about you and a cow?" Lottie asked with a laugh.

"You know J.P., mom. He's a funny kid."

Lottie reached with her free hand to brush a loose strand of hair from Samarie's face. "Not such a kid anymore, though. Neither of you are. When did you get so grown up?" she asked. "I still remember the day you were born."

"Were you scared?" Samarie asked her. "Having me all by yourself?"

Lottie smiled, wistfully. "Every woman's alone when she gives birth, really. But that's not what you're asking me, is it? You're asking me about being a single mother. I already was, remember. You weren't..." she paused, as if considering how to phrase it, "...my first."

"Did that make it easier?" Samarie asked. "That you'd already done it with J.P.?"

"In hindsight," Lottie said, "it seems as if it was a breeze." She reached down and took Samarie's hand. "But why all the questions? You're starting to worry me."

It was Samarie's turn to study her lap, her own hand folded in her mother's. She'd rehearsed a million ways to phrase it. *Mom, I'm in love. Listen, Mom, there's something I need to tell you. Mom, you were a single mother. You understand. Don't you? Won't you?*

"I'm pregnant, Mom."

Lottie took a deep breath and held it in. For a moment Samarie thought her mother was going to explode. "Why would you let that happen?" Lottie exhaled.

"I'm in love. Okay? It just happened."

Lottie's voice clicked up a notch. "We're not talking about love, Samarie. Be in love, fine. Why are you pregnant? What about protection with who-knows what kind of diseases out there now? Where is your head?"

Tears welled up in Samarie. "I thought you would understand. Of all people...."

"Don't go there, Samarie. I was out of high school, I was in college," Lottie interrupted. "It's different. You're seventeen. You'll grow up a lot in the next two years."

Samarie shook her heard. "You haven't even asked me who it is."

"Who is he, sweetie?" Lottie asked, faking a smile. She knew it would be a name she hadn't heard before. Some boy from town. Any of the eligible boys on the Orchard would be like brothers to Samarie. None would be her lover. Lottie understood this much, she'd been there.

Samarie looked up to face her mother, hoping the conversation might begin to swing her way. "Rafi. Rafi Summers," she said. "I know you don't know him, Mom. But he's one of my friends from the outside. I've known him a long time. He's a good guy. I'd be proud for you to meet him."

Lottie dismissed the offer. "Well, how does he feel about this? What are his plans?"

"He doesn't know I'm pregnant, I just found out. But I know he loves me," Samarie said.

"I'm sure he does, Samarie. But I bet this is not part of his life plan right now, to be a parent at this age. It's certainly not your plan."

"You always talk about this plan, Mom. Well, plans change. Yours did."

"Go to bed, Samarie," Lottie said, and she turned away and held still, waiting to hear the door close.

Lottie took a sip of her lukewarm tea. She didn't know what to say to Samarie. She was scared about the baby but something else scared her more. Samarie had grown up. Lottie would have to stop telling her lies. But the truth was ugly.

Walter had decided he couldn't risk flying, or taking his own car. But they'd likely be monitoring the rental facilities, too. In the end, he drove across the river to Council Bluffs and bought an ancient Ford half-ton from an implement dealer who'd advertised it on Craigslist. They agreed he could leave the Iowa plates on until his registration was completed.

He'd moved everything from his company SUV to the car as the dealer watched, then assured him he'd be back to pick up his own car as soon as he could round up a second driver.

The dealer waved as he drove off. Maybe he knew he'd never see Walter again. Maybe he saw things like this every day.

Now, the evening sun was shining right in Walter's eyes. He hadn't slept at all the night before, and when he left Council Bluffs, he headed back across the river and then across Nebraska on sheer adrenaline. He could stop for the night in Sidney, he thought. He'd register under the dealer's name. He had enough of a head start that it would take them a few days to put it all together.

At least, he hoped so.

Walter checked the rearview mirror again. Mid-week on this stretch of I-80, after the 76 branched southwest toward Denver, only a few other cars, pickups, and semis shared the road. If someone were tailing him, Walter would have noticed. Especially because he kept obsessively checking his mirror. Jesus, he was tired.

In the deepening shadows, it looked as if giant rabbits were strolling leisurely across the interstate. Walter needed to stop before he swerved to avoid one. A green highway sign loomed toward him: SIDNEY, 10. Hallelujah and amen, Walter thought. He could already imagine how good that anonymous mattress was going to feel.

But once he'd checked into the roadside motel and eaten the pizza he'd ordered from the delivery card on the nightstand, Walter realized he was too hopped up to sleep. He turned on the television that covered most of the wall opposite the bed and started surfing channels. Cooking. Reality show. Talking heads. Travelogue. Charlie Winters. Parasailing...

He flipped back a channel. There she was again. Charlie Winters. As if she hadn't aged a day. What would she think when Walter Conroy showed up on her doorstep? Would she boot him down the stairs? Probably, he thought.

Walter had been part of the team that had prepared the offensive against Charlie. It was his first assignment as VP of communications, and if he had failed, it would have been his final assignment.

His memory of the meeting that changed his career trajectory was vivid and came back to him as he sat in front of the half-eaten pizza.

It had been a Friday and his mind was beginning to wander off to his weekend plans when Errol Foster walked into Walter's closet-sized office and shut the door. Foster was not known to visit the building where Veregro's scientists or mid-managers worked. Walter and his colleagues had only seen Errol on company brochures and newsletters. And, of course, in the news.

Errol glanced around the office as if someone might be hiding in the corner and then finally settled his eyes on Walter.

"Walter Conrad?" Errol asked.

Walter nodded and Errol smiled as he seated himself.

"Your name keeps popping up, Conrad. I'm hearing good things about you," Errol said, without bothering to introduce himself.

"That's flattering, sir," Walter said. "I'm here to do my best and I'm proud…"

Errol interrupted. "You're a good people person, excellent communicator and a hell of a nice guy," Errol said, holding up a finger for each of his recited bullet points. "That's what I'm told."

"Thank you, sir!" Walter said, trying not to giggle. Errol Foster had more power than the governor of the state, and here he was blowing compliments at Walter Conrad. It was surreal.

"Bob wants to offer you a promotion: VP of communications. And he's asking me to sign off on it," Errol said with a widened smile. "But there's one problem." Errol paused for a moment. "I'm concerned you might be too nice."

Walter wondered if this was some sort of cruel trick that Errol liked to play. What does one say when they are complimented for being a hell of a nice guy and then chastised for it no more than ten seconds later? He was speechless and relieved when Errol starting talking again.

"VP of communications can't be done by someone too nice, Conrad. Is there a glamorous side to it? Sure. That's the easy part. When everyone sings our praises for feeding the world. But I'm sure you know, Conrad, Veregro has to take a lot of shit from a lot of people," Errol said, nodding his head with a frown. "Politicians, reporters trying to make a name, farmers stuck in the twentieth century, there's no end to the shit. And sometimes we have to sling it back. It can get ugly. It's not a good environment for someone, you know, too nice."

Walter nodded.

"This Angell Farm incident is a prime example," Errol continued. "It puts us in the press for all the wrong reasons. That leads to lawsuits. And investigations. It slows the company down. It costs money. That means people starve. You get it?"

Walter was still nodding. "I get it, Mr. Foster. Veregro's work is too important to be slowed down."

"That's right. So we have to do the right thing," Errol said, as he stood up. "We'll have to go public that Charlie Winters fucked up. You know Charlie?" Errol asked.

"Yeah, I know who she is," Walter answered. Everyone in the company knew who Charlie was.

"It's a shame, really. I like Charlie. She's done some really good work here. But the bottom line is she acted alone with Angell Farm. And she didn't follow protocol," Errol said, finger raised for emphasis. "I'm not the best to articulate it all. That's the job of the VP of communications. You look at all the pieces and figure out our statement. You get the heat off of Veregro. Anything less is failure and people starve. It's a huge responsibility, Conrad. Can you handle it?"

Conrad nodded. "Yes sir."

It took Walter exactly five days to secure his new career and crucify Charlie Winters in the process. Three days looking at the pieces, two days writing the press release. In the back of his mind he wondered: was Charlie just a scapegoat or had she really acted alone and broken protocol, as Errol said? On the surface, it hadn't looked much different than any other negotiation that took place with other farms, with the exception of John Angell's reaction.

Rightly or wrongly, the media took the bait. They focused all their attention on Charlie, as if she were responsible for the entire thing. The press releases Walter wrote suggested it was Charlie who'd misled John Angell with false hopes before sticking him with a subpoena to show up at a hearing, where he'd not only lose the land that had been in his family for generations, but demanded he destroy the seed he'd spent the previous ten years cultivating.

It didn't hurt—or, in Charlie's case, help—that she was gorgeous, smart, and charismatic. The media ate her up. The fact she was the bad guy made the story that much better.

In the story Walter wrote and the media regurgitated, with one hand Charlie Winters had handed John Angell the licensing documents, while with the other she assured him of Veregro's beneficent intentions in the world. Being a part of Veregro's Grand Plan meant John Angell could make a difference in the world, as well as leave his family a legacy they'd someday thank him for. His farm could be at the forefront of banishing world hunger forever. All he had to do was sign right there, on the dotted line.

John Angell liked Charlie. Walter had been there. He could tell. But the farmer hadn't bought into her idealistic company line. He'd argued with her, friendly at first, but, as the days passed, more angrily. It wasn't about the world, he'd insist. It was about one man, and his family. It was about their way of life. A way of life. His way of life.

Charlie hadn't seen it coming. Well, neither had Walter, truth be told. But by the time Charlie put the final squeeze on, in the same way the company had put the squeeze on every other farmer whose fields had been tested and found to contain Veregro's patented seed, John Angell had decided he wasn't going to sign a settlement. He wasn't going to destroy the farm he, and his father and grandfather before him, had labored over. He wasn't going to claim responsibility for cross-pollination that had contaminated his fields. They were still his fields. They were still his seeds. If he signed those papers, Veregro would cast a shadow over everything he'd ever done, tainting not just the past, but the future.

John Angell was obstinate. But he was more than that. He, too, had served in the third Gulf War, and, like far too many who came back from that war, his nightmares had never ended. Maybe that was why, one evening, he drugged his wife and children and then, once they'd fallen asleep, shot them in their beds, one by one.

When Charlie Winters got there the next morning to pick up the signed agreement, what she found instead were the lifeless bodies of John Angell's wife and three daughters, as well as that of John Angell himself. Only the youngest, two-year-old John Paul Angell survived, brain damaged, but alive.

The Veregro brass had been relieved to be able to close the book on the whole affair once Charlie took the fall, and Walter had gotten so caught up in his new job—and then his new marriage—that he'd forgotten all about it, until now.

With a groan, Walter pushed himself off the bed and retrieved his tablet, then did a search for John Paul Angell. But all he found was the same old story, the one he'd written, and would relive, again and again, knowing he'd gotten off scot-free while Charlie Winters had taken the fall.

TEN

ERROL'S TABLET WOKE HIM before 7. He groaned as he rolled toward it, realizing at the same time that Alice had already gotten up. That was odd.

The tablet's face came to life as soon as he touched it. The incoming call alert read Bob Howard.

Errol clicked connect. "Did you forget about the time difference?" he growled at the as-always meticulously groomed CEO.

Bob shrugged. "Didn't figure you'd want to wait to hear this. Conroy's gone."

Errol sat up, swung his legs out of the bed, the better to focus on the tablet. "What do you mean, gone?"

"Flown the coop. Hit the road. Hell, Errol. Call it whatever you want. Craig was parked outside his house but he must have slipped out some back way Craig missed."

"For what we pay him, Craig shouldn't miss an albino in a snowstorm."

Bob sniffed a laugh. "That's true. But we both know how smart Conroy is. It's why we put him where he is in the first place. I figured he'd put all those questions of his behind him once he dumped Charlie Winters."

Charlie Winters. Even now, all these years later, Errol couldn't quite get that long tall drink of water out of his system. It had ended between them as quickly as it had started. Charlie had only initiated it in a last-ditch attempt to save her job. Which she couldn't do, of course. That book was written before the ink even dried on Farmer Angell's obituary.

Charlie Winters was back in the news. Errol had seen a story about her hippie commune somewhere in California just the night before. The only picture they could find of her was her official Veregro headshot from eighteen years ago. Errol wondered if she'd aged well. Those hippie women usually went to shit fast, despite their supposedly healthy lifestyles.

Healthy lifestyles. Charlie Winters. Walter Conroy. Errol looked at his screen. Bob Howard was saying something to someone off to his side. Probably Marcia. "Howard? You there?"

"I'm here. You went Deep Space Nine, so I had Marcia bring me another cup of coffee. I figured you were making one of your legendary connections."

It was Errol's turn to laugh. "Oh, I made a connection all right. One we should have figured out before we called Craig. Conroy's on his way back to Charlie Winters."

Bob's coffee cup stopped halfway to his lips. "No. Really? You think so?"

"Think about it, Howard. She's got that healthy hippie commune; Conroy's got a stick up his ass about the company's supposedly less-than-healthy products. He knows we're watching him. He sees Charlie Winters on the news, same as we did, and realizes she's his best hope."

Bob nodded, thinking. "He probably already figured out Randy didn't die of the disease. Randy's tablet was wiped clean but I'll bet you a three-martini lunch he zapped its contents to Conroy before Craig got to him." He tapped a few keys, searching for the location of Charlie's commune. "Bolinas, California. Never heard of it. Have you?"

Errol nodded. "Marin County. North of San Francisco."

"He's not flying. Craig already checked all the airlines. And he's not in his company car. We would have located him on the Nav, if he were."

And just like that, Errol knew where Walter Conroy was. He was driving. He was on I-80, practically a straight shot from Omaha to San Francisco. "When did he disappear?" he asked Bob.

"Yesterday morning. Craig spent a day trying to pick up his trail before letting me know."

"You chew his ass?"

"You bet I did. But let's say you're right, that he left yesterday. That'd put him somewhere in Wyoming, or maybe even Utah or Nevada if he didn't stop."

Errol had opened the FAA weather site while Bob talked, clicked so it was visible on both their screens. "Ha! There's a bear of a spring blizzard in Wyoming right now. Unless he's more of a damned fool than we think he is, Conroy's probably holed up in some roadside motel, catching up on his Zs."

Bob opened the OAG site. "What about airports? Anywhere Craig can fly in before Conroy moves on?"

"The storm's still in the Rockies. Cheyenne would work."

"I'll have Buzz get the plane ready and have Craig meet him at the airport. Plus I'll get an APB out on Walter's company car. Say it was stolen. Offer a reward. That should pop it out of its hole."

Errol was fully awake now. "Maybe I'll head up toward Bolinas myself," he said. "Even if Conroy never gets there, we need to find out what Charlie Winters knows."

Bob didn't say anything for a full ten seconds. "Let's hold that card, Errol. I don't think she knows a thing. If she sees you...Well, you remember what a clever bitch she is, don't you? If Craig... Once Craig takes care of Conroy, Winters won't be any the wiser."

The idea of seeing Charlie Winters again dissolved as quickly as it had materialized. "Keep me posted," Errol told Bob, then severed the connection.

Errol clicked off his screen and listened to the sounds of the big house. No one was in it but him. Errol wondered where the hell Alice was. It wasn't like her to be up—and out—so early.

That damned Sylvia. Next thing he knew, Alice would be talking in Spanish. He'd have to have a talk with Sylvia sooner rather than later.

Walter woke far more slowly than he usually did, unsure of where he was. Once he realized it was the motel room in Sidney, he located the clock. It was already 8 a.m.

Walter practically leapt from the bed. He didn't have time to lie around. He pulled back the curtain to see blue skies. Good. He'd make up for lost time and haul ass today.

He took a quick shower, grabbed a cup of bad coffee and a stiff pastry from the breakfast buffet, then filled up next to the interstate before merging back on. Less than an hour later, he was in Wyoming, traversing a suddenly more barren landscape. No wonder nobody lived here. It looked like the moon.

Up ahead, where Walter knew the mountains ought to be visible, loomed an ominous bank of clouds. Well, it was April. Maybe it was only rain. If Walter remembered right, there were only a couple of high passes. But, while there was always a chance he'd hit a flake or two, Walter could deal with snow. The pickup, he now saw, was 4WD. Lucky. It had never even occurred to him to check, when he'd so hastily bought it the day before.

When he pulled off at a truck stop in Cheyenne, a pair of truckers in the men's room were talking weather while they pissed. "It's a good 'un," the first said. "Had to put the pedal to the metal to stay in front of it. I wouldn't want to be driving into it. No way, no how."

"Guess I'll head down 25 then," the other said. "You know if 70's getting it, too?"

"70, and even 40, down in Albuquerque. You know how these spring storms are. Mean and fast. If I was you, I'd just wait it out here."

The other trucker nodded, backed up a step and zipped his fly. "You're probably right about that. Get the rig serviced, sleep in a real bed for a day. Thanks, Mac. I'll see you at the next pie plate."

The eastbound trucker went to the sink and wet his hands to smooth down his hair. "Whatcha lookin' at?" he asked Walter, in the mirror.

Walter zipped and went to the sink next to him. "I was just listening to you talk about the weather," he said, wishing he weren't wearing the khakis and windbreaker that announced Suit even when he wasn't wearing one. "I'm in an F150 4x4. You think I can push through?"

The trucker squinted at him critically. "You know snow?"

Walter nodded. "I'm from Omaha."

"Yeah, well, this mountain snow ain't like flatlander snow. It blows horizontal, comes at you so thick you can't even see your steering wheel."

"That right?" Walter said.

"Me, I wouldn't drive into it. No way in a semi." The trucker gave Walter another sidewise glance. "In an F150? Maybe, if I was in a hurry. You in a hurry?"

Walter nodded, slowly.

The trucker returned the nod, thoughtfully. "Okay. Take it slow. Don't drive faster than you can see in your headlights, and don't stop. Spring storms come in waves, so right when you can't see a damned thing, you're almost through it. Find a semi to follow, if one's in as damned much of a hurry as you are. You'll probably get a breather by the time you hit the flats in Utah. It's too dry there to hold much snow in the air."

Walter took it all in. "Thanks," he said. "Buy you a cup of coffee?"

The trucker backed toward the exit. "From this dump? I'll pass."

Even though he'd filled the gas tanks in Sidney, Walter decided to top them off. Ford pickups met the EPA minimum requirements for gas mileage, but only by a squeak. If, despite the trucker's advice, he did have to pull off the road for a while, at least he could keep the truck running for the heat. And if he didn't, he could run without a stop until he got to Salt Lake.

If he did...well, Walter would deal with that eventuality if it occurred. Just in case, he bought a Thermos and filled it with coffee, stocked up on snacks after he'd filled up the pickup. The last thing he did before merging back onto the westbound interstate was test the hubs. For a long time, his was the only vehicle on the road.

Lottie woke up drenched with sweat. Flashes of the scene she'd discovered at the Angell house that morning could take her by surprise at any moment, even now. The sound of an opening door, the smell of blood, even an eerie silence, had her once again discovering first, the girls, one at a time, and then John and Christina Angell, in their bed. Blood was everywhere, but despite her gagging and even retching a few times, Lottie felt she deserved what she continued to relive. After all, it was her fault.

John Angell had left a note on the kitchen table addressed to Charlie Winters, knowing she'd be the one to find them. *I'd like to think you haven't completely turned yet,* it read. *Veregro isn't out to save anybody but itself. You can do better. I hope you do. Sincerely yours, John P. Angell*

She still had the note. Nobody, not even the police, knew of its existence. She kept it hidden in her journal and sometimes, on nights like this, she would take it out and read it. Over and over again. *You can do better.* She wished someone had told her that sooner.

ELEVEN

"MR. HOWARD?" Marcia's purr on Bob's tablet was like silk.

"Yes, Marcia?"

"There's a Mick Bruno calling for you."

Bob leaned toward the tablet. "A who?"

Marcia laughed her throaty chuckle. "Mick. Bruno. He says he's an implement dealer in Council Bluffs. Something about a stolen car. I can take a number..."

Bob sat up straight. The APB on Conroy's company car. "No. I'll take it. Thanks, Marcia." He clicked onto the other line. "Howard," he barked as a bulky man in a cramped office appeared on his screen.

"Hello? Oh yeah, hi." Bruno paused, waiting, perhaps, for Bob to say something. Bob let him squirm instead. Never mind that he might have precisely what Bob was looking for. Bob liked making people squirm.

"I'm calling about the stolen car?" Bruno said. Clearly, he'd never outgrown his high school phrasings. Maybe he was younger than he looked.

That happened to those Iowa boys. Built for farming, they went to fat fast when they lived in the city.

"Can you describe it for me?" Bob asked him.

"It's a..." Bruno stopped. "Hey. Shouldn't *you* be describing it to *me*?"

Bob laughed, shortly, then read off a description of Conroy's company SUV, ending with the license number.

Bruno nodded. "That's the one," he said. "He looked like a Suit, not a thief. Never figured he wouldn't be back."

"How'd he leave?" Bob asked him, casually.

"Bought an old Ford half-ton came with an old boy's estate. I was gonna take it over to the auction lot that afternoon if I didn't unload it. Then this guy comes along and says it's just what he was looking for. He didn't look like no F150 man to me, neither."

Bob had already pulled a legal pad toward him to make notes. Most people just used their tablets, anymore, but Bob liked the visceral feeling of scratching words onto paper. He even used a fountain pen, with deep blue ink. "Did you put a temporary plate on it?" he asked.

Bruno shook his head. "Nah. I left my plate on. Told him to toss it once he got his."

Even better, Bob thought. "Can you give it to me?" he prompted Bruno.

"Sure. Iowa—well, I guess you already know that—XYY-84. One of them Save the Family Farm plates."

Bob wrote down the number, began sketching a rectangle around them. "Thanks, Bruno. Appreciate the call."

"Wait. Don't you want to know my 20? Aren't you gonna send someone over to pick up the stolen car? And..." he hesitated a beat, "...is there a reward?"

"Of course, of course," Bob said, already eager to be done with the man. "I'll put you back through to my secretary."

He clicked Bruno off his screen, then buzzed Marcia. "Get this guy's info, have a check for a grand cut to him, and send someone out there to deliver the check and pick up Walter Conroy's company car."

If Marcia had questions, she was smart enough to keep them to herself. "Will do, Mr. Howard," she said.

"And hold my calls," Bob added. It was time to check in with Craig. The man was getting downright slovenly about keeping in touch with the hand that shoveled cash into his bank account.

The trucker had been right about the storm coming in waves, but that didn't make inching through the cotton-balls of snow enveloping Walter's pickup any easier. He'd jumped out just past Laramie to lock the hubs, and that was the last time the pavement had been visible, too. Walter considered checking the Nav to see how far he'd come, but taking his eyes off the road for even an instant could spell disaster.

Instead, he glanced quickly at his speedometer. Thirty miles per hour. It could be a lot worse. He'd crawled through blizzards in Nebraska barely eking out ten. Walter figured he was at least 150 miles past Cheyenne, maybe more. He'd seen a sign for Rock Springs a few minutes earlier, but it was still at least an hour ahead of him at this rate.

Walter hadn't seen another vehicle since Rawlins, and that had been a snowplow. Maybe this was what being dead was like, he thought, alone in a whitewashed world with not even satellite radio for company. He imagined Kylie in such a place. Outgoing Kylie would hate it. Resolutely, Walter chose another heaven for her, filled with friends and chitchat, charity functions and church. Poor Kylie.

He hadn't thought about her since—when?

Ever since Randy Hall had shown up with the news of his dead rat, Walter had been running, first to the Veregro brass, and now, away from them. It occurred to him he might be running the rest of his life. Not a happy thought.

If Kylie were still alive, he'd have kept all this from her. It might have put her at risk. He wondered how he would have handled it. He might not have even told her why he'd left, or where he was going.

Walter refocused on what he could see of the interstate. Slight variations in light distinguished the road from the snowfall, the hood of the pickup from the world beyond it. It was like being in a bubble—

What was that? Out of nowhere, an SUV appeared next to him, slowing when it drew parallel. It must have been traveling far too fast for conditions to come up on him like that. But SUV drivers were often over-confident of their vehicles, and often paid the ultimate price for their cockiness.

This one seemed to be one of those assholes who started to pass then paced the vehicle next to it instead. Walter hated that, even in good weather. Most of the time they were just shitty drivers. But what kind of shitty driver would be out in weather like this?

Walter risked a sidewise glance toward the other vehicle. A big black Escalade. No passenger. Just a driver, a big man whose age Walter couldn't make out. He returned his focus to the interstate in front of him, slowing a tad to remind the guy to finish passing.

Instead, the SUV slowed with him, continuing to keep pace. Walter suddenly felt an ominous foreboding, and at once felt his heart beating in his throat.

He slowed still more, and still the SUV paced him. That was when Walter's thinking seemed to slow, as he considered his options. Up ahead, the sky looked to be lightening a bit. Good. If he could see clearly, Walter could either speed up or pull over.

If he sped up, chances were, if this guy were indeed after him, he'd speed up, too, the better to edge Walter's pickup off one of the precipices Walter knew was there even though he couldn't see them. If Walter pulled over, the guy might walk over and blow him away before putting the pickup in neutral and giving it a push toward a cliff.

Caught between hell and a high place. Walter sniffed an ironic laugh.

Like dueling race cars, the two vehicles continued down the highway at the same pace. But it *was* lightening ahead, Walter could see that now for sure. He touched the accelerator and picked up speed. The SUV did the same.

He could see the edge of the road, as well as the fact that there was nothing beyond it. He must have crested yet another pass without realizing it: They were heading downhill, in long S's engineered to accommodate the interstate's high speeds.

They pulled into a sudden burst of sun. Walter floored it. He didn't even think about it; he just did it. The SUV did the same. Despite the snowpack, Walter drove faster and faster. Sixty. Seventy. Eighty. Ninety.

Up ahead, the road curved left. If Walter were the other guy, he'd go straight instead, just enough to nudge Walter's pickup into whatever lay beyond the precipice.

Walter sped up still more as they approached the curve. One hundred. One hundred ten. He patted the pickup's dash. Not bad for an old gal. Then, just before he began turning into the curve, he slammed on his brakes. The pickup skidded, but it also slowed, fast.

It happened so fast, he barely had time to register it. As he'd suspected, the SUV didn't take the curve but aimed straight for where Walter would have been, if he hadn't braked. But Walter wasn't there. Instead, from ten feet back, he watched the SUV continue unimpeded toward the edge, saw the brake lights and the fishtailing and then, in gut-wrenching slow motion, the SUV disappeared over the edge.

Heart pounding, Walter slowed, then pulled onto the shoulder to a stop. He considered getting out to see what had happened to the other vehicle. But if the drop hadn't been far, if the other driver had survived, it was likely he'd finish what he'd been trying to do when he'd flown off the road.

Walter waited until his heartbeat had slowed before easing back onto the highway. Slowly, he sped up to about fifty, still wary of the snowpack on the unfamiliar road despite the sudden sun.

Around the next curve, he had a view of the mountainside he'd just traversed. About 100 feet below, smoke rose from what remained of a black SUV.

Walter's heart clattered into his throat again, but he didn't slow. It could have been him.

Before him, the clouds were beginning to drift apart. With any luck, he could be in California by morning.

"I just love adventures!" Sylvia, in the passenger seat, told Alice. It was no surprise to Alice, since this particular adventure had been Sylvia's idea.

They'd been watching a video about Big Sur the night before when Sylvia asked her where it was.

Alice had laughed. "Why, it's right here," she'd told Sylvia. "Maybe two, three hundred miles up the coast. It's a full day's drive, though." She pointed to the screen. "Just look at the road."

"Oh, can we go there?" Sylvia cried, once again reminding Alice just how young she was.

"Well, sure," Alice answered.

Sylvia crawled across the big sofa and grabbed Alice's hand. "Can we go there tomorrow?" she begged, her eyes pleading.

Alice had protested that they'd need to get an early start, that Errol was home, that Sylvia might get carsick, but nothing deterred Sylvia once she got a bug in her ear about something. Finally, Alice had decided what the hell. Errol would probably be gone again before dawn anyway. He'd been home so seldom of late, there was no reason for her to stick around just because he was. As for the early start, Sylvia, an early riser, promised to wake her. And as for carsickness,

Alice was the far more likely candidate for that. But she never got carsick if she was the one driving.

So here they were, winding north on Highway 1. They'd stopped for a late breakfast in Cambria, where Sylvia had raved about everything from the blueberry muffins to the coffee. (Should she even be drinking coffee? Alice wondered.) It was mid-afternoon, and Alice realized they'd need to stay somewhere for the night. And she'd have to call Errol. Chances were, he'd either already called out the CHP, or he was hot on their trail himself.

Alice smiled. Errol could be so silly sometimes. Alice didn't need him to leap any tall buildings to prove himself, but Errol was of a generation that still believed some macho posing was necessary. Truth be told, his gallantry was part of what had attracted her to him in the first place. Men her own age were careful to treat women—especially women who owned businesses of their own—as equals. But Alice liked the held-open doors, the daily fresh bouquets, even—especially—Errol's chivalrous performances in bed.

She knew Sylvia wondered why she'd married Errol, but that was none of Sylvia's business. She wished Sylvia would treat him with more respect, though. She told herself that was a generational thing, too, that women in their late teens weren't really women at all yet, and so hadn't yet learned how to treat their elders with respect and deference.

Alice stole a glance at Sylvia, stretched out to her full 5' 8" on the seat next to her, taking in all the beauty of this stretch of California coast. Over the past few months, Alice had learned all about the girl's family and schooling, about her girlfriends and boyfriends and hopes and dreams. Alice herself had had to share very little, Sylvia told her so much.

But almost every time Sylvia told a story of more recent vintage, her face would darken as she went on. Often, she'd interrupt her own tellings before she finished, or digress into another story from when she was much younger rather than continue with what she'd begun.

Something had happened, Alice thought. Something bad, and recent. Something Sylvia hadn't told Errol, and didn't want Alice to know, either.

Well, what better opportunity to find out a deep dark secret than a girlfriends' trip to Big Sur? Given the right conditions, Alice was good at getting her friends to 'fess up. Sylvia would probably be an even easier nut to crack.

TWELVE

THE SUN ROSE BEHIND WALTER as he eased down Donner Pass, and by the time he hit Sacramento, all that was left of the Tule fog were whispers of steam rising from the highway.

Walter had stopped only to gas up and grab fast food that he ate while he drove. After his close call in Wyoming, he wasn't about to let anyone else catch up to him.

Once he was past Sacramento, the highway narrowed, but most of the traffic was headed in the other direction, toward the Capitol, and Walter was able to study the Nav and drive at the same time.

It looked as if his best bet would be to get off in Vallejo. A long series of pontoon bridges crossed the bay where Highway 37 had once been, and from there it looked like a straight shot to the ocean after a short jog on the 101.

He could be there before noon. But what then? Charlie was going to tell him to go to hell and walk away. Or maybe expose him. What then? Walter's foot eased off the pedal. His plan had a gaping hole in it. No, more like a bleeding ulcer. But there was no plan B and he jammed the pedal back to the floor.

Up ahead, Walter saw the sign for his exit, and moved to the right lane. In Vallejo, he filled-up, pissed, and grabbed some food to go before heading west on the reconfigured highway. It was eerie to think all those drowned oil refineries that had been in the news years earlier still lay beneath the bay.

The road couldn't have been straighter if it had been drawn by a CAD program, and Walter's mind wandered, not for the first time, to the question of who'd been trying to kill him, and why. It had to be someone from Veregro, worried he was going to go public with Randy's tests. Now that he'd received Randy's last zap, he could see why. The company hadn't merely slid through FDA requirements; they'd flown over them, skirting any turbulence along the way.

He'd left everything Randy had sent in the encrypted Cloud file, and changed the password to a long phrase he committed to memory. If something happened to him, Randy's notes would die with him.

At least until he got to Charlie's hippie commune where no one had gotten sick....

Jesus. He was thinking in circles, so tired he wished he had toothpicks to prop open his eyes. As he merged north onto the 101, into rush-hour traffic moving more slowly than the joggers next to the highway, he took the opportunity to pour another cup of coffee from his thermos and take another look at the Nav. He'd get off in San Rafael, just over this hill. If all went well, he could be there in an hour.

And then what? That question nagging him over and over. Despite the ramblings of his mind, Walter didn't have a clue.

"We've got him," Bob told Errol.

Errol stopped pacing next to the pool and sat down in front of his tablet. "Craig got him?"

On the screen, Bob shook his head. He was wearing a golf shirt. Errol realized it was Sunday. "Craig's disappeared," Bob said.

"What do you mean, disappeared?"

"After Buzz dropped him off in Cheyenne, he rented an Escalade. He checked in to say it was snowing thicker than his mama's chowder, then called a little while later to say he'd spotted Conroy's pickup up ahead. That was it. Never heard from him after that."

Errol tried to imagine Walter Conroy doing in Craig Stark. The idea alone made him bark half a laugh. "Come on, Bob. Conroy versus Craig? Who's gonna win that one?"

Bob nodded. "Yeah, I know. That's what I thought, too. But when I didn't hear from Craig, I had Rick Fisher contact the authorities in Utah and Nevada, put out a quiet APB on Conroy's vehicle."

"And—?" Errol wished Bob would just get to the point.

"He was spotted in Reno, about 4 a.m."

Errol did some quick math. "He's already gotten to her," he told Bob.

"Or he's close."

Errol stood. "I'm going up there."

Bob patted the air in front of him. "Sit down,

Errol. You can't go. It would muck it all up. And who knows if she'll even see him. There's no way she has warm and fuzzy feelings about Walter Conrad. I'd bet on it."

Errol half-sat. "But if she does? Do we want those two putting their heads together? This could kill us, Howard. This could take Veregro—and, might I remind you, your substantial stock options—down."

Bob laughed. "Relax, Foster. Nobody's taking Veregro down. I'm not gonna let that happen. I figure we give 'em a long leash and see where they want to go, then choke 'em just before they get there."

"You're playing pretty close to the edge," Errol noted. "You prepared to take the fall?"

Bob leaned toward his screen. "Are you threatening me?"

"Who's your majority stockholder, Bob? Who pays for your golf club and your west side mega-house and your Jag and your Oban? Who's calling the shots here, Bob?"

Bob leaned back again, tented his fingers and offered a thin smile. "You unhappy with my performance, Mr. Foster? Because if you are, I can— "

"Of course not," Errol interrupted. "I'm just reminding you who works for whom. *You* don't tell *me* what to do, got that?"

The thin smile remained. "Aye-aye, sir. In that case, however, I humbly suggest that you let me do this my way. I've handled it so far and it's run smoother than a sow's ear."

"Except that Conroy's in California and Craig's AWOL."

Bob chuckled. "Conroy in California's a good thing. Listen. I promoted his girl to take his place. She's already writing up a press release about him hallucinating in his sorrow about his wife's death. Want to hear it?"

It was Errol's turn to chuckle. "Nah. You're a smart guy, Bob. It's why I pay you the big bucks. Just don't forget whose bucks they are."

Bob saluted. Behind him, his wife appeared. "Who are you talking to, Bob? Why, Errol, hello! It's been too long! You'll come for dinner the next time you're in Omaha, yes?"

Errol offered her a far different smile. "Absolutely, Mary. Your pot roast is the stuff of legend."

"And bring Alice," Mary went on, before Bob put a hand on her shoulder.

"I'll let you go," he said to Errol.

"Keep me posted," Errol reminded him before signing off.

Bob Howard had damned well better keep him posted. But just in case, Errol opened a search window. Having a backup plan never hurt.

Alice awoke chilled, opened her eyes to see the fire had gone out. She was in her own cabin, a good thing after what Sylvia had told her the night before. If they'd shared a room, Alice might not have been able to process what Sylvia revealed.

Not that she'd done all that much better on her own. Next to her on the nightstand was the brainstorming diagram she'd drawn just before she'd fallen asleep. She'd hoped to have a dream that cast some light on what she should do, but now she could remember none of them.

Alice sat up, pulling the down comforter around her. The console beside her read 6:11. It also contained a thermostat control. Alice slid one hand out from beneath the comforter and turned on the heat.

It didn't take long for the small cabin to begin to warm, but despite the early hour, Alice was wide awake. With a sigh, she slipped from the bed and turned on the coffee maker before stepping to the floor-to-ceiling window that framed the ocean between a pair of cedars.

Or would have, if the fog weren't so thick. Alice sighed again. She wished Errol were here. Errol would know what to do. But she couldn't even ask him. Not when the question was whether she should even tell him.

After a leisurely dinner at the inn's restaurant, Sylvia had once again lit on the topic of her life in Cusco. Maybe it was the glass of wine Alice had urged on her, having recently read an article that suggested the occasional glass of red actually benefitted pregnant women. Maybe it was the change of venue, or the simple knowledge that they were the only people in the restaurant, that Errol was nowhere nearby, that it was safe.

As usual, Sylvia's words had trailed off halfway through a story about her high school boyfriend, Rafael.

Alice had smiled, and touched her hand. "I think we know each other well enough that you don't need to keep any secrets from me," she said.

Sylvia shook her head. "You will not like this secret," she said.

Alice felt a familiar clutch in her gut, but squeezed Sylvia's hand. "Tell me," she said.

"I became pregnant," Sylvia whispered.

Alice was so surprised she let go of Sylvia's hand. "You have a child already? What? Where?"

Sylvia shook her head, and a single tear tracked down her cheek.

Alice took her hand again. "What happened?" she asked, keeping her voice as gentle as she could despite her growing fear.

Sylvia looked down at their hands. "I—I lose the baby," she said so quietly Alice wasn't sure at first she'd heard her right.

"You miscarried?" she said.

Sylvia nodded.

Alice was too stunned for a moment to respond. Then she began, "Errol would never have…" before trailing off. Sylvia knew the rest of the sentence as well as she did.

Sylvia nodded again. "I lie to him." She met Alice's eyes. "I lie to you. I never mean to. I want to come to America. To California. I want this very much. When Mr. Errol comes to Cusco and I learn what he wants, I know I can come this way. Nobody knows what happens to my—to the baby. Nobody in Cusco. Nobody in America. Nobody anywhere. You are the first person I tell."

Alice was silent. Sylvia was four months along now, showing. No, more than showing: She was glowing. Healthy. The last time they'd visited the ob-gyn, they'd watched their baby—their son—swimming in his amniotic fluid. They'd listened to his heartbeat, as regular as a drumbeat. The doctor had been as delighted as they were.

Sylvia leaned forward. "I did not take the care of myself I do now, Ms. Alice. This time, it will be different. I feel this. I know." What Sylvia did not disclose was that her grandmother had moved from Cusco to Buenos Aires at an early age, and that she and her mother had been raised there. They'd moved back to Cusco when she got pregnant.

"I'll need to tell Errol," Alice said.

Sylvia grabbed Alice's other hand, brought it to her lips. "Oh, Ms. Alice, please, no! Mr. Errol...he will not understand like you do. The men, they are different this way. You know this, Ms. Alice. He will...he will send me away. Then I will lose this baby, too. Please." She turned her head away. "I know I should not tell you," she said.

Alice reached and turned Sylvia's chin back toward her. "No, Sylvia. You were right to tell me. But how can I not tell Errol? He's my husband. This is all his idea. He'll be furious if he finds out."

"Please, Ms. Alice," Sylvia pleaded. "Do not tell him."

Alice pushed her plate away and stood, and Sylvia rose with her. "I need to sleep on it," Alice told her, and they went to their adjacent cabins and said goodnight.

But sleeping on it shed no new light. Each time Alice imagined telling Errol, he'd begun yelling before she even finished. She'd never even had time to explain. If she told him, Sylvia would be gone by nightfall. Along with their baby. Her baby. Errol's.

Errol had set so much store by this child, Alice was afraid Sylvia's past would be the straw that broke his back. The best thing she could do was protect him from himself. Maybe, someday, after the baby was born, maybe she'd tell him. Then, they could laugh about the way she'd chosen to protect him from himself. Or maybe she'd forget all about it.

Alice returned to the window, but the fog continued to shroud the trees. Despite the heat that filled the cabin, she shivered. As soon as Sylvia got up, she decided, they'd head up to Carmel, then cut across and drive home on the 101. The cold and gloom of Big Sur were the last thing Alice needed on a morning like this.

Thirteen

LONG BEFORE HE REACHED THE TURNOFF, it was clear to Walter that he wasn't the only one looking for Charlie Winters—or her commune. Both sides of the tsunami-ruined state highway were lined with pickups and campers that had become homes away from home for their occupants. Some people had set up tents behind their rigs. Others napped beneath the shade of RV awnings. A few camp-spots sported signs, painted with messages like "Share the Health" and "Open the Garden." All were oriented northwest. Walter figured his destination lay in that direction.

The turnoff that led to the Orchard didn't have a sign, but the crowd thickened at its apex and Walter took the turn. As he eased his now-dusty pickup along the narrow two-track, Walter recalled the last news story he'd seen, what seemed like years ago but was actually only two nights before, in Sidney. The talking heads had begun to call these people "pilgrims." They were seeking the Orchard's secrets, but the Orchies had quarantined themselves and weren't letting anyone in—or out.

Well, hell, Walter thought as he eased the pickup slowly between the vehicles parked to either side. While he'd entertained the possibility that Charlie might not see him because of...well, call them old times, he hadn't thought about the quarantine. Who could blame her for trying to protect her commune members?

But Walter needed to see her—and she needed to see him. They needed to share their information, to put their heads together and begin to fix what Veregro—with their unwitting but nonetheless complicit assistance—had broken. They both owed it to the world they'd helped to break.

About a hundred yards uphill, the road ended at a wide metal gate, the kind that was strung between barbed-wire fences at ranch entrances. The gate was closed. Three men stood on the other side, arms crossed.

Walter drove as far as he could, then stopped in the middle of the lane and got out. He didn't bother to lock the pickup. There were too many people around for that to matter.

"You can't park there," called a young man on a lounge chair. He was brushing some kind of dog that he held in his lap.

Walter thought about asking him why not. He decided instead not to respond and headed up toward the gate.

The crowd clearly knew its boundaries. By the time Walter was thirty feet from the gate, he was crossing an unoccupied no man's land.

He didn't get much closer. "Hold it right there," the man posted at the center of the gate called.

Tall. Rangy. Weather-beaten. Walter pegged him in his mid-50s. Handsome in that rugged way that some women liked. He wondered if Charlie were one of those women.

Walter stopped, waved a greeting. "I'm looking for Charlie Winters," he said.

"You're a bit late. Charlie's been dead about fifteen years now, God rest his soul," the man said.

"I'm not talking about Charles. I'm looking for his daughter," Walter said.

"I can't help you," the man said.

"Tell her Walter Conroy's at the gate. She'll remember me. I promise," he said. "We used to work together at Veregro."

Now there was a silence. The other two men turned toward Walter in unison, and then all three simply stared. They all knew of Veregro.

"Look," Walter continued. "She's all over the news. Don't you guys watch the news?"

"Lottie doesn't want to see you," the man said at last. "You're on her Do Not Call list."

Walter took a deep breath. Dust itched at the back of his throat. "It doesn't have to be Charlie," he said.

"Lottie," said the man.

"Lottie," Walter felt his mouth form the word. It might be his ticket in.

"Well, then, who *are* you here for? You ain't making a whole lotta sense," the black man on the right said. There was a laugh in his voice Walter liked, and he turned toward him.

"If you'll tell Charli—if you'll tell Lottie..." Walter stopped. They weren't going to let him in. They knew who he was: the Veregro brass who sicced the media on her, left her alone to face the inquest, the sheriff, even the Senate committees. He was surprised these guys hadn't already pointed shotguns in his face, although it occurred to him they didn't own shotguns. The peaceable kingdom of Charlie—Lottie—Winters. It made sense.

"I know," Walter said. "I'm probably the last person...Lottie wants to talk to. In the world. But for the sake of your community, please. Let me talk to somebody. It's important."

"That's it? That's all you got?" the head man asked. He turned to each of the men at his side. "Whaddaya think, Shooey? Are you moved, Rev?"

Walter didn't rise to his bait. Instead, he went on. "Listen. I have evidence that shows your community's health and MODS...the disease...are closely related."

Now the men were silent, watching him. Waiting.

"I know what's causing the disease," Walter told them. "And I know why you're not getting it. Let me in. Let me help. Please. For humanity."

Walter's words seemed to disappear into the still air. The men looked at each other. They looked again at him.

"I'm not sick," Walter said. "You couldn't catch MODS from me even if I had it. It's not airborne, it's not in our saliva, it's not like that."

"Then what is it?" a female voice asked. A young woman stepped out from between two oaks. Walter watched the woman walking toward him through the trees. He knew that walk. He knew that figure. Charlie Winters.

"My God! Charlie?" Walter said as he gazed at the beauty that had appeared from the trees, that he had last seen eighteen years ago.

The girl laughed. "I'm Samarie." She turned serious. "How do you know that MODS isn't contagious?"

"I'm Walter Conrad. I worked for Veregro, so I can..."

"I know who you are, Mr. Conrad," Samarie said as she folded her arms.

"Then you know I'm a scientist," Walter said, as he reached into his bag and brought out his tablet. "It's all in here. Please, let me show you what I know."

Samarie stared at Walter for a moment before she nodded to Pete. "Open the gate."

Todd knocked on the door of the weather-beaten cottage where Rafi Summers lived with his mother, then turned to study the slough while he waited for someone to answer. Once home to hundreds of different breeds of waterfowl, the slough, befouled by trash and flotsam, was now lorded over only by raucous seagulls. It was still pretty, though, and, if you didn't know its history, calming. But Todd knew how many houses lay under its now-wider footprint. Not to mention how many bodies, both human and bird.

He turned back to the door and knocked again, his fist sounding dull on the weathered wood.

"Hang on," came a voice from inside. Rafi's.

A moment later Rafi appeared behind the screen, T-shirt and jeans hanging easily on his teenaged frame. "Hey, Todd," Rafi said, opening the screen door for Todd to come in. "What's up?"

"Same old," Todd said. "Just thought I'd come by and say hey."

Rafi gestured toward a sagging sofa and Todd sat down. "I'm getting myself coffee," Rafi said. "You want some?"

"Absolutely," Todd said.

Rafi brought him a large chipped mug and settled into a chair across from him. "Is everything okay? How's your mom?" he asked.

Todd took a sip of the coffee. It was good. "My mom is doing pretty bad, Raf. We're just watching her waste away. Doctors aren't doing shit. I'm just glad to be out of the house right now. What's the point, ya know? There's nothing I can do".

Rafi leaned toward him, shaking his head slowly. "Hang out here as long as you want. I'm sorry about your mom." Rafi reached into his pocket and pulled out a small Ziplock bag. "Hey, you know what goes great with coffee? Some Purple OG," Rafi said holding up the baggie with a joint inside. "Let's take this party outside."

They migrated to the weather-beaten plastic chairs on the front patio. Rafi lit the joint, took a large hit and passed it to Todd.

Rafi filled the patio with a large cloud of smoke. "I heard on the news about some experimental drug. Are they giving that to your mom?" Rafi asked.

"Shit, we don't have the money for that," Todd said. "That's for rich people. Not us."

Rafi shook his head, and the joint passed between them a few more times.

"How's Samarie?" Todd asked.

"Good question," Rafi said. "I haven't seen her because of the stupid quarantine."

"They still worried it's contagious?" Todd asked.

"I guess Samarie's mom is paranoid or something. They don't really keep up much with the news," Rafi said.

"Well, they must know something. How come no one there is getting sick? You gotta wonder: Is the quarantine to keep the disease out or to keep some secret from escaping?"

"I never thought about it," Rafi said. "You're living proof that it's not contagious. You're not sick, and you're around the disease every day."

"Yeah, I guess I am," Todd said. "Maybe we should present me as evidence. Maybe we can get inside. That would be great for both of us, right?"

"Todd, you're a genius," Rafi said as he stubbed out the joint. "Finish your coffee, we're going to the Orchard."

Pete stepped toward the gate, lifted the heavy chain around the fencepost and pushed the gate open just wide enough for Walter to step through. Walter looked behind him, expecting the crowd to have surged at this sudden change, but they were all simply watching. Waiting. Quietly hoping they'd be next.

He stepped through the gap and Pete pulled the gate closed again. The tall man stepped forward and looped the chain back over the post. "You sure?" he asked Samarie, his eyes on Walter.

"I'm sure," Samarie said. She looked at Walter, studying his face for a long moment, then glanced at the pilgrims on their doorstep.

Samarie and Pete led Walter through the gate and down the path that led to the bungalow Samarie shared with her mom and her brother. Walter was nervous, like he was being led up a mountainside to be tossed into the open mouth of a volcano. But as they passed magnificent fruit trees, overflowing blueberry bushes, zinnias in full bloom and alive with bees dancing from flower to flower, a calming sensation passed over him. Maybe this was paradise. Maybe everything would be all right, Walter told himself.

"What about the quarantine, Samarie?" Pete pleaded as they walked. "I think this is a bad idea. I'm going on record with that."

"It's not contagious," Walter stressed with a head shake.

They reached the porch of the bungalow. It was a small structure made of brick and wood with stone steps leading up the patio with the front door wide open. Samarie stopped and turned to Walter. "Remember, it's Lottie," she said. "Whatever you do, don't call her Charlie."

Fourteen

STANDING ON THE BUNGALOW PORCH, Walter watched Charlie through the window. She was carefully considering several small jars of herbs. The past eighteen years had hardly written themselves onto her face. She looked just as he remembered. Still the girl from the picture on the news.

Walter knew the years had not been as kind to him. Gray at his temples, thinning hair and the godawful slump of his shoulders from all those years behind a desk. Walter stood up straighter and waited on the porch while Samarie and Pete went inside.

"Samarie, is that you?" Lottie called out without glancing up from the jar she held.

"It's me and Pete, mom. And a surprise visitor," Samarie said, swallowing the lump in her throat.

Lottie registered Samarie's warning tone and looked up, stood up straight and put down the jar of tea leaves. "What do you mean a surprise visitor?" Lottie asked, coming around the kitchen counter and toward the front door.

"Mom, don't get mad. He's here to tell us about the disease. He knows what's causing it. And that it's not contagious," Samarie said as Lottie pushed past her and stormed through the front door.

"Walter. Conroy. Errol's yes man. On my porch." Lottie said, nodding. "Well, that is a surprise."

Walter held up his hands. "I know we've had our differences but you've got to hear me out," he said.

"I don't need to do a goddamn thing," Lottie said. "This is my world you stepped into."

"Samarie's right. It's not contagious," Walter repeated. "I can show you. It's all here." He reached into his bag and pulled out his tablet.

Lottie pointed her finger toward the open door. Walter and the others went inside, and Lottie bolted the door behind them.

"Dammit, Samarie," Pete said under his breath.

"Mom, just listen to him. We can't catch it. I have friends in town with sick family, but they're not all catching it from each other. Please listen," Samarie pleaded.

"Pete, you care to explain this?" Lottie said with her hands on her hips. Pete glared at Samarie until she spoke up.

"I told Pete to let him in, Mom. I made him do it. It's not his fault."

"I am truly mystified by this. Do either of you realize he's probably been sent here on Veregro business?" Lottie asked, looking from Samarie to Pete and back again.

"That not true, Charlie," Walter said.

"Don't you ever, ever call me that," Lottie yelled.

"Lottie. Lottie. I'm sorry." Walter shook his head at his slip up.

Lottie turned to Pete. "Did you search him? Is he wearing a wire? Is he armed? Did you follow any of our protocol at all? One iota, Pete?"

Pete walked over to Walter and motioned for him to raise his hands. "Let's get this over with." Pete patted Walter down. He removed Walter's keys, wallet and phone from his jacket pocket and dropped them on the kitchen table. He untucked Walter's shirt and double-checked for any electronic surveillance. "He's clean," Pete announced.

"Ha. That'll be the day." Lottie pointed to the kitchen table. "Walter, sit. Samarie, you go to your room and wait for me. Pete, outside on the porch and don't let anyone near here."

"What about J.P.?" Pete asked.

"If he comes home you send him on an errand. And keep your distance. God knows what's been brought in here now."

Lottie took a seat opposite Walter at her small kitchen table, tucked beneath a window that looked out into scattered fruit trees that still wore spring blossoms. "I'm not sure what you expect to find on your field trip here, Walter, but you're not going to find it. What's going to happen is this: I'm going to call the police, tell them that you're trespassing here and we conducted a citizen's arrest. And then the police are going to take you away."

Walter let out a sigh. "I don't blame you, Lottie. I drove 1,500 miles before I realized how stupid I was to come here."

"I'm sure Errol's had you do lots of stupid things, Walter."

"I swear to you he has nothing to do with this. I mean, he's part of the reason I'm here, but he didn't send me. Errol would like me dead right about now."

"Well, maybe I've misjudged Errol then," Lottie said.

"Look, I know you hate me, Charlie. Shit! Lottie, I meant Lottie. I don't blame you for directing your anger at me."

Lottie stood up from the table and headed toward her phone. "I think this visit is over, Walter."

Walter stood, too. "Veregro is causing this disease, and Errol knows it. And he knows I know it. That's why he wants me dead. I'm not asking you to believe me but I'm asking you to look at what I've got. Decide for yourself."

Lottie stepped away from the phone, then walked over to the stove and put a kettle on. She slid Walter's tablet over to him as she sat down. "You've got thirty minutes."

"Geez, there's a lot of people," Rafi said. Todd was easing his Jeep between the cars and trucks that lined the road to the Orchard. "'Share the Health,'" he read. "Hey, that's pretty clever."

As they got closer to the entrance, the crowd began to take over the road. There were tents and makeshift campsites all around with people living out of their cars and trucks.

"This is unreal," Rafi said.

"There's a lot of people who wanna know what's going on inside this place," Todd said, as he pulled the Jeep over to the side of the dirt road. "We'll walk the rest of the way."

Rafi was snaking through the thickening crowd, and Todd had to trot to keep up with him. About ten feet from the gate, Rafi burst through into an empty space. The crowd was unusually respectful, keeping their distance from the gate. Looking and waiting for something. A messiah.

Rafi recognized the two men at the gate.

"Hey, Rafi," the white one said. "What's up?"

Rafi walked up to the gate and leaned on it, casually. The men didn't stop him. "Hey, Rev. I've been out of touch with Samarie. I heard there was a quarantine so I wanted to come see for myself. Any chance I can talk to her?"

Todd hung back in the crowd and strained to hear what the men were saying to Rafi. The black man laughed softly. "Young love," he said. "It don't give up."

Rafi turned to him. "Can I?" he asked.

"I can't leave my post, Rafi," Rev said. "Either can Shooey. Look at this place."

Rafi nodded, then turned and scanned the crowd. "Yeah, I see. It's pretty crazy."

"We'll let her know you was here," Shooey said. "But it's gonna be a while 'for we can."

"Tell her I'll be waiting," Rafi said, pushing away from the fence. He walked back over to Todd.

"Young love," Shooey laughed again.

"Well, no way they're letting us in," Rafi said. "I need to talk to Samarie. It's our only shot."

"So what do we do now?" Todd asked.

"We wait for Samarie."

FIFTEEN

WALTER PRESSED HIS THUMB to the bottom right corner of his tablet. Lottie watched as a rendering of Walter's thumbprint emerged as a pulsating glow that quickly faded and the unmistakable sound of a latch clicked, unlocking the device.

"Pass me the tablet," Lottie told him.

"It's encrypted," Walter said. "It will take about five minutes to compile the file into something readable. While it's doing that, let me give you an overview of what I know."

"Well, get on with it then," Lottie said.

"Okay. Over the last two decades the Veregro seed has been modified, many times. Could we make it more impervious to pests, to pesticides, weather fluctuations, invasive indigenous plants? Could it pollinate easier, could we make it survive with less water to fight a drought? And so on. You get the idea. The research never stops. But we always, always kept controls in place. Tests, to be sure the seed was still safe, for the environment, for consumers. The last seed I was personally involved in raised some major red flags after it was treated with Veresate. Our test subjects didn't do well.

"You mean the rats?" Lottie asked.

"Yes, the rats. The first two generations handled the food without any issues. No signs at all. But the third generation, that's where it fell apart. First it was infertility in the females. Then we started to see organ failure in both the males and females."

"What about the fourth generation?" Lottie asked.

"There was no fourth generation," Walter said. "So I reported my findings. That's it. The seed I was testing was experimental. It wasn't the seed we were distributing. In a lab environment, we see such things. We reformulate, we retest. No big deal. It's how science works."

"So it's genetic. That's why it's not contagious." Lottie stated for confirmation.

"Roughly, yeah," Walter said.

"But Veregro rolled the seed out anyway," she said.

"Either that, or a new formula just as bad," Walter said. "But I only learned that recently."

"How so?" Lottie asked.

"A scientist named Randy. You'll learn all about him in a moment, his report is in here," Walter said, gesturing to the tablet. "Randy's most recent tests showed the same thing as mine did nearly two decades before. The difference is that my tests were with experimental seed. And Randy was testing inventory that he grabbed out of a bag on a farm visit in Camarillo."

"So Errol knows that you know. But you've known for twenty years so what's changed?" Lottie asked.

"I didn't know for twenty years. I thought the strain of seed was shelved. It wasn't the first time we tested something that was never produced for the public. A project ends, a new one begins. That was our cycle," Walter said. "Can I trouble you for some more tea?" Lottie got up to refill their mugs and heard the tablet chime.

"Okay, let me wrap it up. The file is ready," Walter gestured at his tablet. "My tests, and ones done more recently. This epidemic is not contagious. It's worse. It's the food, Charlie. It's Veregro."

"Don't call me Charlie," Lottie said as she pulled the tablet toward her.

Samarie sat with folded knees on her bedroom floor, ear pressed against the door. She couldn't make out exactly what Walter was saying, but her mother hadn't thrown him out. That was progress. Suddenly the talking stopped. She must be reading Walter's report, Samarie thought.

She began to worry that her mother would remain unconvinced and that any moment she would hear one of the kitchen table chairs slide across the floor. And her mother would start to yell, telling Walter to get the hell out. No, she would be yelling at Pete, telling him to get Walter the hell out. The Orchard would remain under quarantine and she and Rafi would still be separated while their child grew inside her. Samarie went to her bed and cried herself to sleep.

"Mind if I lay down on your couch for a little bit while you're reading. I've hardly slept in days."

"Knock yourself out, Walter," Lottie said.

Lottie looked down at Walter's tablet. As much as she wanted not to deal with it, the Plexiglas device drew her like a hypnotist. Damn Walter Conroy. Why had he brought this to *her*? Surely there were others in better positions to help him state his case.

Lottie touched the tablet and the file he'd unencrypted appeared. Even though she'd never used a tablet, Lottie quickly realized that it worked much the same way as her last touch-screen laptop. She tapped the bottom of the screen and saw the document was 120 pages long. She sighed, then began to scroll.

The document didn't have a title. It wasn't even organized as a report ought to be. Instead, it appeared a variety of files had been stuffed into this one very quickly, without regard for page breaks, headers, pagination, anything. Graphs overlapped test statistics. Notes flickered with highlighted misspellings. Whoever had compiled this had done so hurriedly.

Randy, the lab tech, Lottie recalled. The one who'd supposedly died of the disease. Whether it was the disease or not, he'd clearly known his end was near. Lottie shuddered. Then she returned to the first page, and began to read.

A file headed "Summary of WC Testing 2012" appeared to be someone's notes on studies in which several generations of hamsters were fed genetically modified seed, some treated with the pesticide Veresate. *Nothing evident in first generation studies*, it read, *but second generation shows lower growth rates + hamsters reach maturity later.*

WC. Lottie looked up from the tablet. Walter Conroy? She'd never thought of Walter as the cowboy type, but she couldn't recall anyone else in the Veregro lab when she'd worked for the company with those initials. She skimmed for a few pages, until she came to two graphs showing the results of tests with rats.

They were meticulous graphs, clearly the work of someone who labored as long over his presentation as he did his testing. Someone who created graphs like these would work methodically and slowly, consider each piece of evidence on its own before allowing it into the larger fabric. Whoever created these graphs was someone whose conclusions Lottie knew she'd trust.

Blue lines recorded the birth rates for first generation rats, green lines the second, red lines the third. In the first graph, the colored lines remained the same through three generations. The control group, fed just the seed.

In the second graph, the green lines were drawn far below the blue lines. And the red lines all redlined— straight at zero. There were no births in the third generation.

Lottie read through the graphs' key. It was straightforward, confirming what the graphs showed. By the third generation, rats fed GMOs treated with Veresate were infertile.

Farmer Angell had been right.

Lottie read on, learning about a European study in which hair grew inside the mouths of rabbits fed treated GMOs. She scanned a newsfeed about infant mice; 50 percent of those whose mothers had been fed treated GMOs died. She read the findings of a

Russian scientific team, which documented that, in young male rats fed treated GMOs, one testicle turned blue.

Lottie's horror grew as she turned the pages. What had they done? They were killing their children, often before they were even born.

Two hours later, Lottie realized she could hardly see anymore. The last of dusk had turned the trees outside her window into the indistinct shapes she'd always found comforting. Only now, the familiar branches seemed to reach toward her, menacing.

Now that she'd read the file, she couldn't do nothing. Damn Walter Conroy. He'd known that, and he'd drawn her back in, as if she were a dumb hungry fish who'd never learned what lay inside the bait.

No. Not dumb. Lottie was older and wiser now. She wouldn't let anyone fool her again. Especially not Walter Conroy. She'd help him, yes—how could she not after what she'd just read? But she'd keep her eyes open, this time. Wide open.

Samarie woke when she heard a kitchen chair slide across the bamboo floor. No yelling followed, though. Instead she heard footsteps heading toward her bedroom, then a knock.

"Samarie," Lottie said, poking her head inside the bedroom. "Can I come in?"

"Is 'no' an option?" Samarie asked, waking from her nap.

"No," Lottie said as she came in and closed the door.

"So, is Walter gone? Did you have him arrested?"

"He's sleeping on the couch," Lottie said.

"So you read his report," Samarie said, sitting up on the bed. "Is there any good news at all?"

"First things first, Samarie," Lottie said as she sat down on the bed. "You used poor judgment letting Walter in here without talking to me first. Pete has a soft spot for you. You know that, and you took advantage of him."

"It was just that, Walter looked so desperate, saying he had important information for us, and I knew you wouldn't listen unless I brought him in," Samarie said.

"I know why you did it, Samarie."

"So now what?" Samarie asked.

"You're going to apologize to Pete. And you're going to promise me that you will bring issues like this to me and only me."

"I'm sorry, Mom. I promise."

"Okay. Now get yourself up and get some tea. It's going to be a long night for you."

"You're going to tell me about the report?" Samarie asked.

"You're going to read it yourself," Lottie said.

"Are you serious? I have homework tonight."

"This is your homework tonight," Lottie said.

Samarie slid off her bed and slipped her feet into her sandals before shuffling off to the living room, where Walter was snoring loudly on the couch.

"Before you get started, go apologize to Pete and give him this," Lottie handed Samarie a folded sheet of paper. "It's a shopping list."

Samarie unfolded the paper. "What is this stuff?"

"Prenatal vitamins. It's for you and the baby."

SIXTEEN

ALICE CLOSED THE TABLET on her big tidy white desk and spun around to take in the view out her office window. She'd headquartered her business in Santa Barbara long before she met Errol and moved with him into the Montecito hills. Now she was grateful it was here, a place for her to avoid that same home.

She wasn't avoiding Errol; God knew he wasn't there enough to be avoided, anyway. No. She was avoiding Sylvia. Ever since the young woman had told her she'd miscarried, Alice had tried to calm the conflicting emotions that coursed through her.

She was angry, yes—angry that Errol hadn't vetted his choice for their surrogate properly—but she was also worried. She was worried their own child might meet the same fate, of course. But she was also worried about Sylvia—absurd, considering how deeply Sylvia had deceived them.

But she liked Sylvia. Still liked her, in spite of this breach. The problem was that Alice wasn't accustomed to dealing with emotional issues. Coddled as a girl, now a trust-funder (and married to one of the world's wealthiest men), she'd been sheltered from squalid scenes her entire life. Now, it seemed, they confronted her every way she turned.

Alice considered the ocean through the palms that framed her third-story window. What she needed to do, she decided, was get back to work. Yes, she was here, in her office, had been for a week now. But now she needed to be her old self again, conjure a new product line out of thin air, fire up the troops with her vision, then lose herself for the months it took to develop and promote it. By then the baby would be born, Sylvia would be gone, and Alice wouldn't have to think about any of it any more.

She spun back to her desk, opened her tablet and began scrolling idly through her address book. Business cards appeared before her briefly, then disappeared, each nonetheless calling up a face to go with a name.

Elena Medina.

Alice tapped the screen to stop the scroll. *Elena Medina. Community Gardener. Ojai.*

Alice remembered Elena Medina. She'd given a talk about organic gardening, and afterward they'd chatted briefly about the possibility of an organic line of cosmetics under Alice's brand. And that had been how many years before?

It didn't matter. This was exactly what Alice was looking for. She tapped the number on the card and a moment later was connected to Elena Medina's voicemail. *Please leave a message and I'll call you back,* said the smiling dark-haired holo.

"Hi Elena. It's Alice Foster, president of the Beauty Is line of cosmetics. I'm hoping to start an organic line, and remembered our conversation. Can you call me back?"

After she disconnected, Alice opened her doodle-pad and began to sketch. By the time Elena Medina called her back, she'd come up with a name and design prototype for the new line. But, best of all, Elena remembered her, and invited her to come to her garden. They set up a time when the garden would be in full bloom, in May.

Alice had forgotten all about Sylvia.

Lottie and Samarie were making breakfast in the kitchen when Walter woke himself up with a loud snort. He looked at his watch. "Wow, have I really been asleep for..."

"Yeah, fifteen hours," Lottie said.

"Good morning, Mr. Conroy," Samarie said.

"The bathroom's over there," Lottie said, pointing. "Get yourself cleaned up and come have some breakfast."

Walter shuffled off to the bathroom while Lottie and Samarie set the table with fruit, eggs, bacon and tea. "Samarie, don't get too comfortable with Walter," Lottie warned.

Walter returned, sat down and poured himself some tea. The way Lottie was frowning at him suggested he wouldn't be staying long. But how could she be angry at him? They were both after the same thing, weren't they? The tablet still sat on the table where Walter had left it before he passed out on the couch. He decided to break the silence.

"So you read it?" he asked.

"Of course I read it," she said. "I didn't sleep all night because of what's in that tablet."

Relief coursed through Walter like a calming breeze. "So you're going to help me?"

Lottie pitched her anger directly at him. "I can't help you, Walter."

"You're the only one who can help, Lottie. Look, I know you're angry about the past, but we need to go after Veregro," he told her. "Not each other."

"If you think this is about what you did all those years ago, you're wrong. I'm not going to help you because this isn't my mess. It's your mess. It's Veregro's mess. My job now is to take care of my family and this Orchard."

Walter started to speak but Lottie interrupted him.

"You sat on these tests for years and did nothing. You sang Veregro's praises. You're worse than Nero, fiddling while Rome burned. You were the one doing the burning."

"That's not true. I reported my findings, every detail."

"And then what?"

"And then I was promoted. I wasn't in charge of the lab anymore. But I never doubted that they took my work seriously. My reports."

"What a coincidence."

"What are you saying, Lottie?"

"Oh please. Errol bought your silence, Walter. That's why you were promoted. A scientist promoted to VP of communications. You never wondered why?"

"I'm gonna go finish getting ready for school," Samarie said as she got up from the table and stepped away from the tense conversation. Walter got up from his chair and paced back and forth, silent for a moment.

"Sure, I wondered why. Why me for this job? Well, why the hell not? I worked my ass off for Veregro. Half their patents are my work. Mine. So they recognized me and moved me up the chain. Yeah, I took it," Walter gazed out the window. "That first year after the promotion. Probably the best year of my life. Kylie and I got married, moved into a house. I earned it." He turned back to Lottie.

"It's called denial, Walter. I can only imagine the guilt you're feeling but you can't drag me into it. I have my own issues to deal with," Lottie said.

"Denial?" Walter laughed. "Did you say denial? I just showed you that the world is being poisoned and you think this is my issue? You're not going to escape this. How long before your crops are contaminated by Veregro? It's just a strong breeze away. Will you do something then? People are dying, Lottie. Everywhere. You would know that if you left this place once in a while."

"Thanks for sharing the report, Walter. Really. I don't keep up with news on the outside like I probably should so it's important that I know all this. But I think it's time you go now. Go back to your wife, your family."

"My wife is dead, Lottie. And our baby."

There was a stunned silence. Finally, Lottie said, "I'm sorry." It was all she could think to say. She felt for him, she did, but her mind wandered.

Idiopathic Infertility Syndrome. Samarie, J.P., third generation. What if? Had they actually tested their own crops? Were they all contaminated, too? Maybe just a generation behind. What if? "Please sit down, Walter."

Walter sat down and stared at the black screen of his tablet. His shoulders slumped.

"I wish I could help, Walter, but I wouldn't know how to help you. I've got to take care of the people here. If the outside world isn't going to make it, maybe those of us here can. I need to care about those closest to me first."

"Lottie, listen to me. You are the key," Walter said. "No one here is sick. Not with MODS or anything else. You can help people, teach a new generation how to survive this. We'll expose Veregro, until someone listens."

"Walter, slow down. This is not something I can do right now. My name has been in the news recently and I don't like the publicity."

"I've seen the story. It's great. We can capitalize on it."

"It's not great. The spin in that story is Angell Farm. I will not go through anything like that again." Lottie bit at her thumb nail and looked away from Walter. "Why is this happening? This fucking reporter asking questions, now Angell Farm is in the news again."

"I have an idea," Samarie said. Both Walter and Lottie looked over to her. They hadn't noticed she had come back into the room. "You should invite him here."

"What are you talking about, Samarie?" Lottie said.

"Simon Keller, the reporter. He's been blogging about the Orchard, too. He thinks this place is a great mystery." She laughed a little. "So let him in."

"It could work," Walter said. "Offer him an exclusive if he drops Angell Farm from his story. The real story is this place, the lifestyle, why no one is sick here."

"Exactly," Samarie confirmed. "And we'll be helping people."

Walter got up and kissed Samarie on the head. "You raised a genius here, Lottie."

Lottie sat silent for a moment until a thin smile rose on her face. "Samarie, tell Pete to call a dinner meeting for tonight. We've got business to discuss."

Everyone noticed the stranger sitting next to Lottie in the dining hall and she knew she would have to start the meeting immediately. Lottie stood up and gave everyone a moment to settle in.

"Everyone, I'd like you to meet Walter Conroy," Lottie said, motioning for Walter to stand. Then she swept her outstretched arm from left to right. "Walter, this is everyone." There was a small chuckle around the room. "Well, it's a quorum at least. Thank you all for coming. Let me tell you a bit about Walter."

After Lottie told them who Walter was, Walter gave an overview of his research and findings, linking Veregro's products to MODS. The room was silent. Veregro. Veresate. Dead rats. Idiopathic Infertility Syndrome. Multiple Organ Dysfunction Syndrome. Dying people. The fourth generation.

The silence broke when Walter explained why the disease was not contagious. There was an audible sigh of relief in the room. He wanted to end on an upbeat, so he wrapped it up. "I know what I've told you tonight is disturbing, but there is hope. All of you here are so very fortunate. You have a chance to help people now, on your own terms and I truly hope you will embrace it."

Walter sat down and Lottie gave him a quick embrace and a smile. "Spoken like a true pro, Walter."

Lottie rose and quieted the room. "So I think we should talk about lifting the quarantine." The teenagers in the hall cheered. "I think I spot some 'yes' votes in that corner of the room," Lottie joked. "But I have another proposition to discuss as well."

It was two hours before everyone who wanted to speak had their turn. The vote to lift the quarantine was nearly unanimous. The vote to invite Simon Keller had to be conducted twice. Several hands went up and down in the midst of counting. That was a no-no and always meant a re-vote. It was a polarizing topic and a large segment of the Orchard distrusted the media. Period. It was a close vote, but in the end, Simon was voted in.

Rafi and Todd were sitting on a log not far from the Orchard's front gate. The last few nights they had slept in Todd's car, but when they weren't asleep, they blended into the crowd and tried to kill time, just like the other pilgrims who had come to the Orchard.

"We got anything left to eat?" Rafi asked Todd.

"We're down to water and granola bars," Todd said, fishing through his cooler.

"Jesus, one more granola bar and I'm gonna puke. Are there any blueberry left?"

"Nope. Try apricot." Todd handed Rafi a granola bar and a bottle of water. "We'll go into town tomorrow and stock up."

Rafi nodded. "How 'bout one more game before we turn in?"

"You got it," Todd said, and they started to set up the game pieces on the backgammon board between them.

Rafi suddenly looked up and twisted his head toward the gate. "I think I heard someone calling my name."

Todd froze and listened. "You might be hallucinating." They continued to set up the game, but then they both froze.

"Slow down, Samarie," Pete yelled.

Rafi and Todd got up and ran to the front gate. There was no mistake now; a female voice was calling Rafi's name. They got to the gate and there was Samarie, clutching the padlock.

"You're here," Samarie said when she saw Rafi.

"I said I'd wait here and I did. Didn't Rev or Shooey tell you I was here?"

Samarie just shook her head and began to cry. Rev and Shooey looked at each other.

"You didn't tell her he was out here?" Rev asked Shooey.

"You said you was gonna," Shooey replied. Rev just shook his head.

Pete was leaning over with his hands on his knees, breathing hard. "Pete, you okay?" Rev asked.

"I'm all right. I was trying to keep up with this

one." He nodded toward Samarie. "Flying like a bat out of hell. The quarantine's been lifted."

"Hallelujah," Rev said.

"We got permission to let Rafi in." Pete was fumbling with his keys but Rafi climbed over the fence in two strides and swept Samarie into his arms.

"We can't leave Todd out there," Samarie said to Pete.

"Just hop over, Todd," Pete said with a wave of his hand. Some of the other Pilgrims saw the boys hop over the gate and started approaching. "Pete flicked on his flashlight and shined it on the approaching men. "Don't even think about it. These two here got clearance. Official business." Pete saw others watching the scene from their campsites. "The gates will officially open tomorrow, folks. As long as everyone remains orderly, you'll all get a chance to come in. Now try to get some rest." But there was no chance of that. Everyone was talking and would continue until daylight.

Samarie turned to Todd and gave him a hug. "I'm so sorry about your mom, Todd."

"Thanks, Samarie." Todd cast his eyes down. He felt better when it was off his mind. He perked up as if remembering something. "Hey, I'd love to say hi to Willie and Bren. Are they around, too?"

"They're headed to the rec room." She pointed to a well-lit path lined by trees. "Follow this walkway. It will be the first building on the left. Just listen for the music," Samarie said with a laugh, pulling Rafi close to her. "We'll catch up with you guys in a bit."

Samarie led Rafi to a grass field that was filled with Maidenhair trees and horses sleeping in the leaves. They lay in each other's arms, nestled in the pile of drifted leaves from the previous autumn. Samarie kept touching Rafi's face, to assure herself she wasn't dreaming, that he was really here, that this was really happening. Rafi seemed half-asleep. That was fine. She could study him without him studying her back.

He had such an unusual face. His skin was the color of the leaves they lay in, his long wavy hair nearly black. His eyes were black, too, and were framed by long lashes any girl would covet. That was part of what had drawn Samarie to him in the first place: She'd never seen black eyes before. Of course it didn't hurt that he was also the best looking guy she'd ever seen.

Would the baby look like him or like her? Samarie conjured a little girl half herself, half Rafi, then laughed at the image she'd created.

Rafi opened his eyes, then smiled lazily and kissed her. "What's so funny?" he asked her.

"I was just imagining what our baby will look like," she told him.

Rafi sat up. "Our baby," he said.

Oh, shit. They were so...connected, and she had known for so long now, it seemed impossible that Rafi didn't yet know. She touched his cheek again, then cut right to it. "I'm pregnant, Rafi."

Rafi laughed. Samarie loved his laugh. "What are the odds?" he said. "Do you know how many women are trying like anything to get pregnant? And then, you and me do it one time and— really? You're pregnant?"

Samarie nodded, smiling.

Rafi stood and pulled her up, then began spinning her round and round in some joyful dance. "Then we're never going to be apart again!" he cried.

Samarie laughed, spinning in a circle. "I'm so happy!" she cried. "I'm the happiest girl in the world!"

Rafi stopped spinning and pulled her back into the leaves, kissed her and held her close. "I am so lucky," he told her. "I love you so much."

"I love you, too," Samarie said, and began to cry. She'd never understood that before, how people cried when they were happy. But now she did.

Simon sat in his car across the street from a house where he was watching a middle-aged woman preparing dinner. She stood at a kitchen island flipping around on her tablet and taking small sips of white wine. Every few moments she turned around, stirred something on the stove, tasted it, then went back to the tablet and wine. This continued until dinner was ready. And it was the same as always. Dinner for one. She brought her tablet to the table along with her food.

Simon was looking at a tablet in his lap. On it was a celebrity gossip website. It was flipping from story to story, seemingly by itself. Various searches for Amazon and Google appeared in the search bar, but Simon never touched it. It was all synchronized with the woman's actions. Simon had hacked into her wireless connection, and was able to watch her closely with the magnification built into his iGlasses. Her websurfing was boring him silly.

Green text popped up in the lens of his iGlasses: BLOCKED NUMBER. Unknown Caller. Simon rolled up his window, then answered.

"This is Simon Keller." Before Lottie could respond, the computer in Simon's iGlasses decoded the blocked information. Marin County, CA, (415) 555-1882, Charlotte Winters.

"Simon, this is Lottie Winters calling. Lottie," she emphasized. "I hope I'm not calling too late."

"Not at all. I'm always working. I gotta say I'm surprised to hear from you."

"Something you need to understand about me is that I'm very protective of my family. And the Orchard is my family. This place is sacred to us. It's our Garden of Eden. We're wary of strangers."

"I came on a little strong. I get that. But I'm a reporter. I ask questions. I look for stories."

"You're trying to dig up dirt on Angell Farm. We can't be friends if that's your agenda. I went through hell over that incident once and I'm not going back. And I will fight anyone who tries to take me there."

"So, why'd you call?"

"Because I like what you had to say about the Orchard in your story. If you didn't cloud it up with Angell Farm, you might be a decent spokesperson for us."

"Okay, I'm listening," Simon said as he switched off the tablet with the feed from inside the woman's house.

"If you phase Angell Farm out of your story, I'm willing to offer you an exclusive on the Orchard. You can come live here for six months, see what we do, how we do it. Blog from the inside. We won't talk to other reporters. You'll be our voice."

"That sounds like an offer too good to refuse."

"It is," Lottie said. "And if you don't take it, I'm calling someone else."

"No need. I'm in," Simon said.

"It's the right choice, Simon. I think this will be good for both of us," Lottie said. "I've got to warn you though, we're a bit unsophisticated here when it comes to technology."

"Yeah, I remember," Simon said. "You don't need to worry about any of that. I'll bring everything we need. I'm a bit of an expert with that stuff."

INSIDE THE GARDEN OF EDEN
A blog by Simon Keller

I have just arrived at Winter Farms here in Marin County, California. If you come around the surrounding area asking for directions to Winter Farms, chances are you will be met with a blank stare, because everyone around here, especially the inhabitants, just call it the Orchard. Some locals will joke that you can follow your nose to the Orchard. It could be the smell of ripe berries or tangy citrus, sometimes it's the flowers and once in a while, the manure.

But the first thing I notice isn't a smell, it's all the bugs. The preying mantises cross the dirt roads like they own the place. Ladybugs, blue beetles, yellow beetles; they decorate the plants like moving polka dots. Then there are these little green buggers with transparent wings. They seem to be everywhere and several try to take up residence in my left ear.

Arachnoids are technically not insects so let me please give the spiders their proper respect. I'm not particularly bothered by bugs, especially if they stay out of my ears, but some of the spiders here are downright intimidating. I see webs that I swear would keep me warm at night. So I ask Pete about all the bugs.

Pete is a man of many hats here at the Orchard. For starters, he oversees the care for everything that grows here, everything produced. While he leads me to my cabin, I ask him if they do anything here to control all these bugs. I thought most of them are bad for crops. That's why most farms use pesticides or GMO's, isn't it?

Pete is tickled by my questions. Well actually he laughs in my face but I am okay with it because he teaches me something I didn't know. The bugs and spiders I see are eating the destructive bugs that would consume the crops. So the bugs control the bugs. Nature's pesticide. Safe for the environment. Unless you happen to be an aphid or a cutworm, or a flea beetle, or some other harmful bugger. No GMOs or chemical pesticides have ever been used at the Orchard.

I'm about to ask Pete another question when he puts a finger to his lips and points to his ear. Universal language for shut up and listen. So I do. I hear a humming sound, like electrical cables buzzing nearby, but I don't see any cables or wires. The sound seems to be coming from a field of tall sunflowers so I walk closer. I stop when I see the bees. Hundreds, thousands, maybe a hundred thousand?

I can't tell you how many bees it takes to bring that sunflower field abuzz, but I do know a little bit about bees. I know that they are an essential link in the chain of food production for mankind. Without bees working their inadvertant pollination magic, we would have no apples, avocados, coffee, lemons, strawberries, or watermelons or... I'll stop now because the list is really long.

I also know that year after year bees are dying off quicker than their previous generation. Their numbers have been plummeting drastically since about 2005. Twenty-five years of research hasn't gotten us closer to explaining precisely why they are dying in massive numbers. Some scientists say it is toxic pesticides and GMOs, others point to global warming. Maybe both are right, but here at the Orchard I would say bees are thriving. As are the crops and the cattle, and from what I hear, the people.

If you've read this far, I want to thank you for joining me on this journey. In recent months the Orchard has caught the eyes and ears of the public. This small community seems to be untouched by the MODS and infertility epidemic that is sweeping across the world at alarming rates. Locals close by have long known that no one gets sick at the Orchard, long before MODS. Is it a fluke? No one knows for sure but I hope I'll find out by spending some time here and I am grateful for the invitation sent by Lottie Winters, heiress of the Orchard. Through this blog I will share with you what goes on here. Maybe there are lessons for all of us. So welcome... Inside the Garden of Eden.

Seventeen

Now that Walter Conroy had surfaced at the Orchard, Errol pondered his next step. The campaign to discredit Conroy had gone off without a hitch, thanks to Conroy's newly promoted assistant's inside knowledge about the man. Newsfeeds headed *Grieving Widower Blames Employer for Wife's Death* had been replaced by *Veregro Regrets Former Communications VP's Lapses.*

Lapses. Errol chuckled every time he saw the word. The very sound of it suggested a man whose mind had gone blank. Still, now that Walter Conroy and Charlie Winters were teaming up, Errol wasn't about to drop his guard, clever campaign or no. Being proactive was always the best course. But it was also a method that Errol particularly relished.

Ever since Bob Howard had relinquished control of what he called The Walter Conroy Problem to Errol, Errol had been free to do things his way, without any interference from the—to his mind—over-cautious CEO. Every time he'd told Howard what he knew would happen next, Howard would counter that it wouldn't hurt to wait a day or two and see if the wind shifted so they couldn't smell the offal. Errol wasn't just happy to be free of Howard; he was delighted not to have to listen to any more of his folksy aphorisms.

So. What would Conroy and Winters do now? Errol thought about what he would do, given the chance, and allowed himself a brief diversion into the joys of Charlie Winters's body. If only she'd learned to keep her big mouth shut, things might have turned out very differently for her. And for Errol.

But Charlie Winters thought she knew everything, and her undoing had been the shortcuts she'd taken because she'd figured no one else would notice her when she was moving so quickly. Her mistake was that Errol could think as fast—even faster—than she could. Errol had always been one step ahead of Charlie. The Angell Farm incident nearly tripped him up. But even that eventually had worked to Errol's advantage, in the end.

So, let's say Conroy convinced Winters there was something to that lab tech's tests, that the treated seed was causing the disease. Errol pulled up the report Conroy had tried to share with the board, using the tablet where his girl had uploaded all his data. How had he left his own tablet behind? Errol hoped he wasn't starting to lose it. His father had disappeared into Alzheimer's when he was only a few years older than Errol was now. Sure, Errol religiously took the preventative Veregro had trademarked, but that didn't mean a niggle of worry didn't edge in from time to time.

Hell, he was unfocused today. To get back on track, he opened a document file and typed in a filename: *Endgame*. Errol liked giving his files names only he'd understand the significance of.

One finger tapping a key lightly, Errol considered what <u>he</u> would do if he were Conroy and Winters. For starters, he'd blog what he thought he knew. *Blog*, he typed. Conroy would know what to do to help a blog go viral. But...Errol paused, thinking, then grinned when the answer came to him. If Veregro announced it was sponsoring a charitable food distribution program at the same time Conroy was trying to discredit it, Conroy would be written off as a crank.

Still grinning, Errol clicked on Walter's former assistant's tablet number. The new Veregro VP of communications had already proved her worth. Now let's see what she could do with this little gem.

By early May, Simon's blog was gaining a sizable audience. Three hundred thousand unique hits daily. Lottie had nominated the rec room to be the official workspace. She checked in daily to see what was going on and was a little annoyed at how attracted she found herself to the technology. It seemed like magic. Powerful magic.

Rafi, Todd and J.P. hung around the rec room a lot and Simon found them distracting, so he put them to work. They were thrilled to have something to do and free access to the Internet. Walter was showing Lottie around to explain what Simon's recruits were doing.

"We can view all the major newsfeeds simultaneously on these screens," he explained, gesturing to the wall above the long table where Rafi and Simon sat, typing on tablets. J.P. was patiently using a graphics program to create a platform for the Orchard.

Because she was watching J.P., Lottie didn't respond to Walter right away. If it hadn't been for all of this, they might never have known J.P. not only could work with the design program, but had a knack for it. He was so engrossed, in fact, he didn't even notice her looking over his shoulder as he worked. In the past, he'd been tuned to her proximity as if he possessed a special radar.

Lottie turned back to Walter. "What are Rafi and Simon doing?" she asked.

"Rafi runs keyword searches. Anytime a new feed appears, he sees it instantly."

"What are the keywords?"

Walter tapped his own tablet and opened a document. "The Orchard, of course. But also Veregro. GMOs. Veresate. MODS. IIS." He looked up as one of the screens on the wall shifted scene. "I'll be damned," he said as a middle-aged blonde appeared on the screen.

"What?" Lottie asked. "What is it?"

Walter pointed. "It's Katrina. My assistant. Only now she's me. VP of communications for Veregro. Todd? Turn up the sound on 3, would you?"

The audio increased even though Lottie hadn't seen Todd do anything. "...access to food is at an all-time low," the woman on the screen read from prepared notes. "But we at Veregro are committed to combating food insecurity..."

"Ha!" Walter barked. Lottie turned to him. "'Food insecurity,'" he said. "It's a good brand. She's doing a good job." He shook his head, as if torn between pride and disappointment.

Lottie turned back to the screen without responding. "...no one should go hungry when there's enough for everyone. That's why Veregro will give you not just seed, but food grown from Veresate-protected seed, if you need it, when you need it. Please visit our website..."

"Jesus!" Walter exploded. "They're going to spread it even more quickly. How could they? It's...it's..."

"It's business as usual," Lottie said quietly. "Did you really think they'd wait to see what we did before striking? They're being proactive. Just as they were when I was out there peddling licenses." She stopped for a moment, looked at J.P. "They don't care about people, Walter. It's all about the money. That's all it is."

Walter shook his head and sank into a chair. "I can't believe I was a part of it for so long."

Lottie put a hand on his shoulder. "You're working to undo it now, Walter," she told him. "That's what's important."

"I think it's time to move on to the second phase, Lottie."

"It's too soon."

"It's not," Walter said. "Veregro is already trying to get the public on their side. They're moving on this. We need to act now, Lottie."

"What do you suggest?" Lottie asked.

Walter touched his tablet back to life and tapped open another file. "Ms. Winters goes to Washington," he said.

Lottie took a step back. "Oh no. Not me."

Walter leaned forward. "It's got to be you, Lottie. It can't be me. I can't go public. They'll either discredit me or...or, I don't know what." He gestured toward Simon and Rafi. "The guys here are too young to establish the credibility we're after. That leaves... you."

"Who on earth am I going to talk to in Washington?" Lottie asked.

"Becky Shaver at the FDA."

"Becky? She used to work with Jim Baker, in R&D, right?"

Walter nodded. "Until she went to the FDA about ten years ago. I have a hunch Becky has no idea of the extent of rubberstamping going on at her agency."

Lottie shook her head. "What if she won't see me?"

"She will," Walter said. "You're the representative of the coalition that's disseminating relevant findings. It's her job to hear what you have to say," Walter said.

It was Lottie's turn to sink into a chair. "Government-speak. God. I thought I'd left it all behind."

She considered raising all the objections she already had. Who would believe her, after all? But she remembered Walter's argument, too, that if she, of all people, was now speaking out against the seed she'd been so hot to license, people would listen. Lottie sat back for a moment, looking across the rec room toward its wide double doors. All she had to do was stand up, cross the room, and push through them. And then....

And then, she'd have to live with herself for the rest of her life, knowing she could have done something, but instead just walked away. With that sobering thought, Lottie scooted her chair closer to Walter's. "Okay," she told him. "Show me the plan."

Eighteen

Simon was in the rec room working on his weekly edition of Inside the Garden of Eden. Todd and Rafi were taking a break and playing ping pong. J.P. was absorbed in his graphics work, tuning everyone and everything out. Simon's focus was broken when his iGlasses began to vibrate on the desk. Incoming call. He popped the glasses on to see who was calling: Errol Foster, Burbank, CA. Holy shit. Simon reached into his bag and fumbled with something in his backpack before he trotted outside to take the call.

Simon cleared his throat. "This is Simon Keller," he said, as the image of Errol Foster appeared in his lenses.

"Simon Keller, this is Errol Foster. Do you know who I am?"

"Everyone knows who you are, sir."

"Hmm. I guess that's true," Errol said with a smile, taking a sip from his tumbler.

Jesus, Simon thought. Was the man drinking whiskey at nine in the morning?

"I called to congratulate you. I've been reading your blog and it's good work," Errol said.

"Thank you, sir. That means a lot coming from you."

"How's your traffic?"

"It's not bad, sir. We're averaging over a quarter-million visitors a day. Trending up on the weekends."

"And advertising?"

"We're in the black. My bills are paid and I'm making some money on top of it, so I can't complain."

"Some money? That's too bad. Someone as talented as you should be making a lot of money."

"Can I quote you on that, sir?" They both laughed.

"Listen, Simon. I've been in the media business a long time. I know how this machine works. The big money comes with wider exposure. You're local. You gotta go national. Hell, worldwide. That's what we do at Foster Media. We find talent and promote it. You should come work for me."

"Wow. This is a lot to think about, sir. I mean, I can't think of a higher compliment. The thing is, my independence in what I do, the stories I choose to pursue. How do I put it? That freedom kinda fuels the fire."

"You know, Simon, that's what Maya King told me the first time she turned me down, and now... she's Maya King." Errol emphasized that with a big smile. "Look, I don't want to control your posts and tell you what to report on. That's not what I do. What I want is to see your posts before they go to print, just like an editor would. Except I don't make changes. Suggestions maybe, I can't help myself sometimes, but no demands on your stories. My job is to make them big. Get you on TV, cross-promote with other media. Go global."

"It sounds great, sir. But there's always a catch, " Simon said. He couldn't help himself, it was too good to be true.

"You're a smart kid. There is always a catch. The catch is about the Orchard. I want more details that don't go to print. What's it like there? What are they doing? What are their plans? Who knows, maybe I can help."

"All due respect, sir. Why not just talk to them and ask?" Simon asked.

"That's a longer story, Simon. I can't talk to them because they think I'm a bad guy." Errol saw the wheels turning in Simon's head, processing. "You just say 'yes' and I put you on the payroll. How does two hundred and fifty thousand sound?"

"Two hundred and fifty thousand a year is a lot of money, sir. But even on my own I think my potential is..."

Errol cut him off. "Two fifty a month, son."

Simon absorbed it. "That sounds like an offer too good to refuse."

Alice had forgotten how much she loved being on the road, even if it was only a short drive like the one to Ojai. Where the ocean had taken over the thin strip of land the 101 had once traversed, a pontoon roadway now floated above the waves. It was almost like flying. Alice could feel the tension she'd been carrying release her neck from its stranglehold.

Just before Ventura, she took the exit for the highway that wound up the hill to Ojai. Little had changed here; an old hippie community of trailers and clapboard houses still hugged both sides of the road after it narrowed to two lanes. As it wound higher, Alice caught brief glimpses of the lake to her left.

When she pulled into Ojai a few minutes later, her Nav indicated she should turn left. Once she had, Alice searched for a parking space. The garden occupied a block in the middle of town. She could walk. It would feel good.

She'd forgotten how much warmer it was here, especially in mid-May. Even though the town was less than ten miles from the coast, its altitude meant the marine layer stayed down below. Alice crossed to the shady side of the street, then set out for an open space a block ahead of her.

The lush block was so thick with growth, it might have hidden a house at its center. An arched entryway made of fruit trees beckoned, and Alice ducked underneath. A well-trodden walkway drew a straight line between raised garden beds. To either side of the row she walked, similar pathways drew long parallels.

Alice peered into one of the beds to her right. Lettuces, of more varieties than she could name. Another was thick with lavender, and the one next to that, thyme. Basil. Oregano. Sage. Rosemary. When was the last time she had inhaled these scents in their native form? Had she ever?

A woman appeared at the end of the row then walked toward her, holding out her arms. "You must be Alice," she said, her voice warm and earthy.

Before Alice could take a step back, the woman reached for her hands, then pulled her into an embrace. She smelled of soap and soil. Coolness and earth. She stepped back, and held Alice at arms' length. "Welcome to my garden," she said. "Actually, to *our* garden. This space belongs to all of us women."

Alice stepped out of her grasp. "Elena," she said with a smile to cover her discomfort at the easy hug.

Elena's smile seemed as permanently etched as the laugh lines next to her eyes. "Before I launch into my usual spiel," she said, "why don't you tell me what you know about the garden already?"

Now Alice smiled genuinely. "I know that it's a community space for women," she said, like a schoolgirl. "That you began with heirloom seed but that everything now self-propagates. You—the women—you're just caretakers."

Elena offered her smiling nod again. "What's been most wonderful is how much the women get from the garden. I don't mean food. I mean renewal. Creative sustenance. Self-faith. Symbiosis."

Without even realizing she was doing so, Alice reached out to run her fingers along the tops of the sage. Elena reached out and ran a sprig between thumb and forefinger. "Do this," she said, holding her fingers out toward Alice.

Alice sniffed. It was the purest thing she'd ever smelled. "I have to have these for my line," she said. "This is what I want."

Another smile from Elena. "Do you want to come here, or do you have a place you can garden back in Santa Barbara?"

Alice shook her head. "No. I mean I want to buy these herbs from you. I'll pay whatever you like."

For the first time, Elena stopped smiling. "They're not for sale," she said.

Disappointment churned through Alice. "But I thought..."

"This garden is for the women," Elena said. "We tend our plots, we reap our harvests, we renew our souls."

Alice re-gathered herself. "So perhaps I could talk to one of the other women..."

Elena interrupted her. "Alice. You're not listening. You need the garden, too. I watched you when you first arrived. I saw how the garden beckoned you. I saw how you couldn't help but touch it."

As if she had no control over it, Alice's hand reached out toward the lavender, ran a sprig between her fingers as Elena had done with the sage, then brought her fingers to her nose. Heaven. It was heaven. "I'm not sure I could come to Ojai all that often," she began.

Elena's face relaxed back into its characteristic smile. "Do you have somewhere you can garden in Santa Barbara?"

Alice considered the land surrounding her and Errol's house in Montecito. There was the rose garden. There were the grapes. There was the broad lawn to either side of the driveway...

She'd never liked that lawn. It screamed, *We have money!* It hollered, *We don't care!* Why not have a section of it torn out and turned into a mesclun and herb garden with a fruit and berry orchard, a place where plants could propagate themselves, and, more important, Alice could propagate *herself*? Elena was right. The herbs were drawing her in.

"How do I begin?" she asked Elena.

Elena smiled. "Come," she said. "I'll give you some cuttings. Tell me what you want, what calls to you. We can collect them together."

"Errol Foster?" The petite blonde on the screen squinted forward, as if she didn't believe her eyes.

Errol smiled. "Becky Shaver. How ya doin', sweetheart?"

Becky Shaver had left Veregro to work for the FDA maybe ten years ago, long enough to have risen through the bureaucratic ranks to become Deputy Assistant Something-or-Other. Errol was counting on her remembering the great reference he'd given her, all those years before.

"I'm good, Errol. God, it's been forever. I mean, I see you on TV—it's so generous of you to be offering all that money to find a cure for IIS—we've gotten lots of proposals in—but I can't remember the last time we saw each other."

Errol now remembered the woman's propensity to jump from topic to topic, only to somehow land back where she'd begun. Good thing, too, as she'd lost him about two leaps in. "You've done good, haven't you, Becky?"

Becky smiled. "Thanks to your leg up, Errol...."

Good, Errol thought, while she nattered on. When she paused for breath, he asked her if she remembered Charlie Winters.

Now she shook her head. "Poor Charlie. Of course I remember her. And talk about coincidences, she just called me the other day. She's coming to Washington this summer and wants to talk to me about something. Should I tell her you said hey?"

Errol nearly rolled his eyes at the woman's naiveté. How had she ever risen through the ranks with that addled brain? "Oh, I suspect Charlie Winters won't want to hear a hey from me," he said.

Becky sniffed a self-deprecating laugh. "Oh, I forgot. Of course. Well, mum's the word. But why are *you* calling, Errol? After all this time?"

Errol paused, considering the best approach. "Actually, it's about Charlie," he said at last.

The banter was gone. Becky sat back, folded her hands on her desk. "Go ahead," she said. The change in demeanor was striking. Errol was almost proud of her.

"You remember Walter Conroy?" he asked.

Becky nodded. "Sure. Of course. How's Walter?"

Errol put on a sympathetic look. "Well, I'm sorry to tell you that Walter's in a bad way. He lost his wife to MODS, Becks, and he took it hard. Real hard."

"I'm so sorry," Becky said. "Poor Walter. But what's that got to do with Charlie Winters?"

Errol rolled a pencil back and forth on his desk. A pencil! When was the last time he'd used one? "Okay, Becks, I'm just gonna give it to you straight. Walter got it into his head that it was Veregro's products that killed his wife. Some crazy lab tech was doing testing on the side, fell in love with his lab rat, and then the rat died. Next thing you know, he's enlisted Walter, right after the poor guy's buried his wife. Convinced him her symptoms and the rat's were identical. Talk about hitting a man when he's down."

Becky's hands remained folded on her desk, although she offered a low nod. "And?"

Was she questioning him, the little bitch? He could pull the rug out from under her faster than she could say *and* in that smarmy little voice. Still, Errol pressed ahead. "We thought we'd convinced Walter it was bullshit. Then he disappeared. Now he's surfaced. At Charlie Winters's Orchard."

Another nod. "And Charlie agreed with the evidence Walter showed her?"

Errol shrugged. "Go figure."

Becky was studying him. "Can Veregro defend itself against that kind of claim?" she asked.

"What the hell, Becks? The FDA approved all our products eons ago. Back when you were still with us. Are you questioning your own agency's methods? Are you questioning the protocols *you* designed?"

A series of rapid blinks indicated that he'd gotten her. "Of course not," Becky said. "Those studies followed protocol to the letter." She paused, studied something on her desk Errol couldn't see. "But Errol, listen. She's on my schedule. If I cancel, I've got to give her a good reason. What do I tell her?"

Errol relaxed. He hadn't just gotten her, he was pulling the strings now. "Why not listen to what she has to say? Find out her strategy. Make her feel heard. And then," he offered her his best smile, "for old times' sake, let me know what her story is. Sound like a plan?"

Becky nodded. She wasn't smiling, but that was okay. He didn't care if she was happy. He just needed her to do his bidding. "Hey, Errol?"

"What, sweetheart?"

"Any progress on the IIS front?"

Errol shook his head. "Not a nibble. Not even the reward hounds. You'd think they'd at least try."

Becky shrugged. "It's a tough one. We haven't had any cures come across the desk here, either."

A zap popped up in the corner of Errol's tablet, distracting him. "Errol?"

"Sorry, sweetheart. Something just came in I need to take care of. Give me a holler after you meet with Charlie, okay?"

"Will do," Becky said.

Lottie had forgotten how hot Washington was in June. Thank god for the air-conditioning, she thought, as she stepped into the vestibule at the FDA. Right on time, she was ushered into the office of Becky Shaver, deputy commissioner of food science. The walls were beige and the office modest, except for the heavy furniture that looked as if it might have grown right out of the floor. "Thank you for seeing me," Lottie said, holding out her hand.

Becky had come around her desk. She didn't merely shake Lottie's hand, but took it in both her own. "Charlie! It's been such a long time," she said. "What have you been up to all these years? You just flat-out disappeared off the face of the earth."

Lottie sat in one of the chairs facing the desk, and Becky returned to her own chair. "I went home," Lottie said. "And then I stayed there. In fact, this is the first time I've left the Orchard—except for trips to town—since I got there."

"Wow," Becky said. "Something must be pretty important for you to come out of your hidey-hole. So tell me, what can I do for you?"

Lottie took the tablet out of her bag and set it on the desk. She still wasn't as easy with the thing as Walter and the boys—Rafi and Simon—were, but she knew her way around. She unlocked the system with her thumbprint, then tapped open the file she'd placed front and center. Then she looked across the desk at Becky. The little blonde was showing her age. That happened to petite women, Lottie thought, wondering for a moment whether it also happened to taller ones like herself. She hadn't, until this moment, thought to look.

Dismissing these thoughts, Lottie turned to Becky. "Before I begin," she said, "can I ask if you've had any family or friends affected by MODS or IIS?"

A shadow passed across Becky's face before she quickly resumed her composure. "Why?" she asked.

"It's just that so many people have lost someone. I wanted to tell you how sorry I was, if you...."

"Is that why you're here?" Becky asked, her tone suddenly sharp. "To play my heartstrings?"

"Of course not," Lottie answered quickly. "It's just that the data I'm about to share with you could help us fight the diseases."

"Really," Becky said. She was all business now.

Lottie wondered what had changed, whom Becky had lost. Clearly, Becky didn't want to go there, so Lottie positioned the tablet so Becky could see it, too. "Let's start with an overview," she said. "As you can see from the data we've gathered..."

Becky interrupted. "Who's 'we'? How many of you are on this project?" she asked.

Lottie actually had to pause and count. Walter. Rafi. Simon. Todd. Herself. She could have included Samarie and J.P., even Bren, but they were just kids, still needing her protection. "Five," she said.

"All from the Orchard?" Becky asked.

"Actually, one of us is an ex-employee of Veregro." Walter would be furious, but Walter wasn't the one sitting here, now, was he?

"An insider," Becky said. She actually sniffed. "With a possible agenda."

Unsure what to make of this, Lottie went to Plan B. She started with Samarie's friend Chrissy, in Bolinas, whose mother had died so suddenly Samarie still got weepy about it. Then, without mentioning her by name, she told Becky about Walter's wife, about their efforts to conceive, and her sudden and rapid death when she was only a month away from delivery.

Becky listened without asking any questions, periodically squinting closer to the photos Lottie opened her on the tablet.

Next, Lottie showed her the graphics J.P. had drawn to show the MODS's geographic distribution. Becky slowed her down here, leaning closer to the display before nodding and sitting back again.

It was time to share the data from Walter and Randy's tests.

Lottie slid the tablet across the desk so the obviously nearsighted Becky could see the results for herself. "These tests were conducted independently, by two different men, eighteen years apart. But their results were identical. The disease incubates for two generations. It strikes the third."

Lottie and Walter had gone back and forth about whether to target Veregro or just the Veresate-treated GMOs Walter and Randy had tested. Lottie had prevailed. Veregro wasn't named; the products were called by their in-house names, KLG21366 and KVB10132.

But of course, Becky had worked for Veregro, too. "Those are Veregro products," she said. "Ones the FDA tested back when you and I were still in the Veregro ranks."

Lottie nodded.

Becky tapped the tablet off and pushed it back toward Lottie. "I appreciate your documentation," she said. "But it's quite a claim you're making. Do you know how many lives will be affected if this becomes public?"

"That's why I'm here," Lottie said. "Veregro has a responsibility to take a serious look at this data, not just brush it under the corporate rug."

Becky folded her hands on the desk. "These products met rigorous FDA standards," she said.

"I understand," said Lottie. "But the FDA didn't do cross-generational testing on treated GMOs. These studies—"

"Studies?" Becky interrupted. "Pet projects, you mean. What I don't get is why *you're* here. Is this revenge, Charlie? Is this about Veregro running you off all those years ago?"

"Of course not! Did you look at these results? Do you think I'd risk everything I've built for myself to come and attack Veregro if I didn't think there was something here?"

Becky tented her hands. "But why don't you go there? To Veregro?"

Lottie paused to contain her growing anger. "My partner tried," she said. "They ran him off." Almost literally, she thought, but didn't say.

Becky shook her head. "But I'm sure Veregro would share your concern about their products' potential link to loss of life."

Bureaucratese, Lottie thought, her hopes sinking. "I agree," she said to Becky. "But apparently, they need a little...encouragement. That's why I'm here. I believe the FDA should ask Veregro to run specific and independent trials on their products."

Becky looked somewhere past Lottie's shoulder. "Of course, this is a long-term issue."

"That's exactly what I'm saying. But the term has come in. We're in the third generation. Now. People are dying. If we don't act now, there won't be a fourth. We don't have time to weave through the bureaucratic maze."

"Might I remind you that I work for what you so cleverly term 'the bureaucratic maze'?"

Shit. "Sorry." Lottie smiled. "You know what I mean."

Becky didn't return the smile. "Let's get down to brass tacks here, Charlie. Who else have you talked to about this?"

Lottie shook her head. "No one," she said. "Obviously, I'm concerned about the ramifications. If this got out, there would be panic. I have another meeting at the EPA. I'm hoping they'll recognize the urgency of this matter and implement a rapid response."

Becky suddenly sat up. "Oh! Who are you talking to over there?"

"Patricia Brouard."

"Wonderful!" Becky's demeanor changed so quickly it was as if she'd traded her professional chill for Midwestern camaraderie. Lottie wasn't fooled.

"Listen, Charlie," Becky said, standing. "Thanks so much for coming in. I appreciate your concern.

We'll certainly discuss whether to approach Veregro. If, in fact, that's what reason calls for in this matter."

Lottie wanted to protest, but Becky had already walked toward the door. Lottie tucked the tablet into her bag and stood as well. "Please call me if I can provide any more information," she told Becky. "And I'll zap you a copy of everything I've just shown you."

"Thank you," Becky said again. Then, finally, she got the woman through the door and shut it after her.

Back at her desk, Becky flipped open her tablet and called Patricia Brouard at the EPA.

Her old friend's face lit up when she saw who it was. "Becky!" Trish and Becky had arrived in the District at the same time. As the saying went, they had history, both professional and personal.

Becky cut straight to the chase. "I hear you have a meeting with Charlie Winters," she said.

Trish opened her calendar. "Why, yes, I do," she answered. "Friday. Another GMO complaint, from the looks of it. You know how often we get those. It's usually from the farmers, trying to find an 'environmental' way to battle Veregro."

"This one's a little different. Do a search. I'll wait."

Trish tapped from her calendar to a news feed. "The Angell's Death. Well, I'll be damned. I'd forgotten all about that."

"Yes, and the Angell's Death just left my office. But listen, Trish. Why don't you let me handle this one? There's no need for us both to waste our time. It's the same story. She's going to come in with one of those complaints you're used to seeing. I've already done the due diligence on it, and can spare you the follow-up."

Trish laughed. "Aren't you the work fairy!"

Becky smiled. "You know me," she said, "Happy to sprinkle my magic dust on your full plate and make some of it disappear for you. Still, you should take the meeting. Hear her out, you know? Why not? I'm on it already. No follow-up required."

"Thanks, Becky. I owe you one."

"You bet you do. Friday after work at the usual place?"

"You're on," Trish said.

Becky disconnected, then pulled up Errol Foster's number. No. First, coffee, then Errol Foster. Let the Big Man sweat a little.

NINETEEN

ERROL TOLD HIS PASADENA HOUSEKEEPER he was heading back to Montecito and went into the Burbank office only long enough to grab his tablet and let Francine, his assistant, know his plans. This morning, she was resplendent in a little bit of a red dress. One of these days, when Errol had some time...

But he didn't have any time, at the moment. He had Francine call ahead to the Burbank airport, where he'd left his jet, and by the time he parked the Mercedes, the jet was fueled up and ready to go. Errol ran his own flight check, of course, pausing to stroke the plane's sleek body. She was a beauty, and she and Errol had been through a lot together. Truth be told, she was his best girl, with him for the long haul.

Once he'd taken off, it was only a half hour to Santa Barbara, and by 10 Errol was winding the Bentley up the hill. He always took enormous pleasure in this drive, especially once he'd turned onto his own road. To the west, a steep drop-off fell into the ravine. To the east, high walls and hedges masked the estates of those who, like him, had not merely made it, but prospered.

When he turned into his driveway, though, Errol was surprised to find the gate wide open. He was still more surprised to find his wide expanse of lawn being torn apart by heavy equipment.

Errol threw the Bentley into park and leapt out. He marched to the person closest to him, a hefty man in khakis, work shirt, and hard hat. Errol grabbed him by the arm and spun him around.

"What the fuck—?" the man said. He was half a head taller than Errol and twice as broad. Errol took a step back despite his anger.

"That's my question precisely," Errol said.

The man studied him. "Where's your hard hat?" he asked.

Errol stepped back into the man's space. "I don't need a fucking hard hat. I own this place. So tell me what the fuck has happened to my lawn."

The man's face relaxed. "Of course. You're Errol Foster. Pleased to meet you." He offered his hand, and Errol shook it, reluctantly. "I'm Victor Sanders. I'm the foreman. Your wife's got some great ideas for what we're gonna do here, Mr. F. Her and Ms. King, I should say."

Errol's gaze followed the man's gesture toward where Alice and Maya stood, far up the slope, wearing hard hats of their own, a half-unrolled site plan held between them. Without another word to Sanders, Errol was back in the Bentley, roaring up the rest of the drive, then skidding to a stop under the portico.

Alice and Maya looked up at his approach, and by the time he hopped out of the car, Alice was trotting toward him, a broad grin on her face. Man, she was gorgeous. Errol felt his head of steam dissipate as quickly as it had arrived, replaced instead by pure lust for his wife. He'd missed her.

By the time Alice reached him, Errol was hers. "Errol! I didn't know you were coming back today!" she said after a warm hello. "This—" she gestured, "was going to be a surprise."

Errol looked out on his ruined lawn, one arm draped around Alice's shoulder. Maya was walking toward them now, wearing a smile as big as Alice's. "Oh, it's a surprise, all right," he told Alice. He held his free hand out to Maya as she reached them. She squeezed it, then offered a quick hug as well.

"Your wife is one brilliant woman," Maya said. "I am so excited about her organic garden, I'm going to do a show about it. Well, not hers," she amended, seeing Errol's glare. "About community gardens. About heirloom plantings. Back to the earth. All of that."

"But not here," Errol said, just in case she hadn't grokked his glare. He was her boss, after all.

Maya shook her head. "Of course not. I feel the same way you do about home as sanctuary."

They turned to watch the dozers and loaders tear up what little was left of what had been a perfectly manicured expanse of Kentucky bluegrass. That lawn had always said *you've arrived* to Errol, but Alice and Maya wouldn't understand that. Well, maybe Maya, who'd worked her way up just as Errol had. But Errol wasn't about to show her—or anyone—that particularly deep chink in his armor. They might be the women closest to him, but they were still...*women*.

Until this moment, he'd had no idea just how important the lawn had been to him. Watching it be destroyed was almost as bad as those miscarriages must have been for Alice. Errol felt his knees begin to buckle, and squeezed Alice's shoulder. Unaware of his sudden weakness, she leaned her head on his shoulder and squeezed her own arm more tightly around his waist. It was just enough to bolster him.

Lottie was halfway through her hotel breakfast on Friday morning when a slender young man in a suit entered the restaurant. Suits were no strangers in this town, but something about him set off her internal alarm. She watched him as he searched the room. Once he spotted her, he made his way to her table. Nice to know her internal radar still worked, even after all these years. Now it prickled the hairs on her neck. Whoever he was, this guy was not bringing good news.

"Charlie Winters," he said.

Lottie didn't respond. Let him squirm a little.

"Mind if I join you?" he asked.

Lottie looked around the restaurant. Half the tables were occupied. Waiters and bussers hustled between the dining room and kitchen. She looked at the Suit and shrugged.

He sat down across from her, then slid a business card across the table.

Rick Fisher
Corporate Counsel
Veregro Industries
Omaha, Nebraska

Lottie looked at him across the table. "God. Did I look that young when I had your job?" she asked.

The kid didn't even crack a smile. So much for breaking the ice with a joke.

Fisher extracted a tablet from his briefcase and tapped it on. "We understand your interest in Veregro has been quite keen of late."

"With good reason," she told him. "I have a report I think a lot of people in Washington will want to see."

"You might reconsider that," Fisher said, with no expression.

"Sorry. Have a nice day, Rick Fisher," she told him, looking down at the card.

Fisher leaned toward her across his tablet, a young slender guy's version of a threatening stance, perhaps. "I've seen Conroy's files," he said. His voice cracked in the middle of the word Conroy.

Lottie stifled a laugh. "Good. I'm glad you have them. It will save me having to send them to you. But let's cut to the chase, shall we, Mr. Fisher? We can swing egos all morning, but truthfully, the test results demand further investigation. Even if you're tied to the doghouse, you can see the shit out back."

Veregro's dog didn't bite her bait. "You and your silent partner are playing with a fire you can't even imagine the size and strength of."

"Apparently you didn't do your homework on me," she told him. "Didn't your keepers brief you on the Angell's Death?"

"Do you want to cause worldwide panic? Do you have any idea what havoc accusations like these will wreak?"

"Why do you think we haven't gone public, Mr. Fisher? We're trying to do things the old-fashioned way. By the book. Tried it lately?"

Fisher tapped his tablet, opening a file. Then he looked up at her. "Your sarcasm is not amusing."

"It's amusing me."

Fisher turned the tablet so it was facing her. "I don't have time to trade *bon mots* with you, Ms. Winters. I suggest you take a look at what I've got here."

Lottie didn't want to look, she really didn't. She'd worked for Veregro. She knew whatever was there was going to change everything. But the thing was right under her nose.

PETITION FOR EMINENT DOMAIN

Lottie skimmed the page, ignoring the lawyer in her that told her to read every word, slowly and carefully.

She looked up at Rick Fisher. "You can't do this," she said. "I own the Orchard, free and clear."

For the first time, Fisher smiled. He had tiny little teeth, like a Venus Fly Trap. "But we can, Ms. Winters. We have every right to petition the appropriate government agencies to further our public service of research in order to feed the world's population. As you're learning, we have many good friends in high places."

Lottie shoved the tablet back across the table toward him. "Errol Foster's behind this, isn't he?"

"No one's, as you so eloquently put it, *behind this*. Veregro isn't steered by a captain. It's a community of people, committed to the greater good."

Lottie slammed her teacup onto the table. "Bullshit. How can you spout the company line without a trace of irony? No, no, don't answer me. I used to do it, too. When I was only a little older than you are, from the looks of you. Have you ever thought about the great engine of Veregro, Fisher? Have you ever seen what a tiny cog you are? You could be replaced like that." She snapped her fingers. "You don't matter."

Fisher tucked the tablet back into his briefcase then sat up again. "Of course I don't. But I care about world hunger. That's why I work for Veregro. But you—you want to bring it all to screeching halt. You'd rather see panicked people on the streets, trampling each other as they grovel for food that doesn't exist."

"Oh, you really drank the Kool-Aid, Fisher," Lottie said.

Fisher shook his head. "I see two little people— an old hippie who's hidden from the real world for eighteen years and a recently bereaved widower whose grief was used by another for his own malign purposes—against the company that's saved the world from hunger. You're wrong. We can't let you do this to the world. Hence the eminent domain." He tapped his briefcase with a forefinger.

Lottie looked at the briefcase, then back at Fisher. He blinked. "When was it filed?" she asked.

Fisher smiled again. "Funny you should ask that. It hasn't been. We know that's your little corner of the world. We don't need it. Whether we take it or not depends entirely on you."

"You're blackmailing me."

Fisher rolled his eyes. "Of course not. I'm persuading you not to go public and to stop all testing.

It's what good lawyers do. Surely you remember that, Ms. Winters."

Lottie waited but he didn't go on. "Is that all?" she asked.

"I take it I have your understanding that if the terms of our offer are not met in full, we will file this petition without warning."

"And?" Lottie asked.

"And we'll be watching." Fisher got up and left.

"Take care, Fisher," Lottie called out. "I'm sure the next time a Veregro lawyer comes to harass me there'll be a different name on this card." She ripped his card in half and threw it to the table. The other diners were silent and trying not to stare. Lottie looked down and cradled her hands around her mug of tea. Then her tablet started ringing.

"This is Lottie Winters."

"Ms. Winters, this is Nancy Farmer, from Patricia Brouard's office." She spoke very slowly and had a look on her face like she was about to say she just burned the cookies. "I'm so sorry to tell you on such short notice, but Ms. Brouard is not going to be able to make your meeting today. A pressing issue has come up and she is going to have to give you a call to reschedule."

"For later today?" Lottie asked.

"Oh, no dear. No time this week. So sorry. Ms. Brouard will get in touch." And Nancy Farmer became a black screen. Washington had shut its doors.

Beneath the grape-covered arbor next to the swimming pool, Alice, Maya, and Sylvia finished their lunch. Sylvia had begun to show since Alice had last spent any time with her. And she'd felt the baby kick, she'd said, dispelling what was left of Alice's concerns. Alice laid her hand on Sylvia's belly. She could swear she felt a kick herself. Her baby! Alice beamed.

After little more than a week at home, Errol had had to jet off to Omaha. Or was it Atlanta? Alice had to laugh at her inability to keep track of his comings and goings. They'd had a particularly lively conjugal visit this time. Now that Alice was wrapped up in work and her new garden, it was hard to remember how depressed she'd been only a few months earlier.

Errol had expressed some regret about the grass, but Alice had laughed it off and that was that. Now though, as they watched the new garden coming to life before them, she told Maya and Sylvia, "Poor Errol. He did love his lawn."

Maya shook her head. "Men and their lawns. I've never understood it. Give me a garden any day. I'd much rather get down and dirty than sit around looking at something that doesn't *do* anything."

Alice giggled. "Oh, Errol likes down and dirty just fine."

Maya wagged a finger at her. "Don't go there unless you intend to tell all," she warned.

Alice finished off her fruit tea, courtesy of Elena, and poured herself some more. "Would you like some, too?" she asked the others, then refilled their glasses when they nodded.

"*Gracias,*" Sylvia said.

Alice turned to her. "You've been awfully quiet, Sylvia. Are you all right?"

Sylvia nodded. "Oh yes. I am fine."

"I hope you'll help us once the garden is finished," Alice said. "Nothing strenuous, of course. But there's plenty to do that's not hard work, and I just can't tell you how good it feels to make things grow."

Sylvia nodded at her belly. "I do know this," she said.

While a statement like that might have sent Alice spiraling the month before, now she simply laughed. She turned to Maya. "You said you had something to tell me," she said.

Maya turned her chair so that she was facing Alice. "Good thing you reminded me. Middle-aged minds, I swear! But listen. Have you ever heard of a place called the Orchard?"

Alice shut her eyes, the better to try to place the name. "Wasn't it on the newsfeeds back in the spring? A bunch of hippies up in the Bay Area, right?"

Maya nodded. "I wouldn't call them hippies, exactly. They're more like Elena. Propagating heirlooms from seed and then letting nature take its course. I heard they eat only fruits, organic lettuces, meat. And they're thriving. Utterly thriving."

Alice turned and winked at Sylvia. "I hear a show coming," she said.

Maya nodded, her face lighting up. "It's what I told you about last week, doing a show about organic gardening. Only this is more. We'll talk about self-healing. We'll talk about the back-to-earth movement. This is tailor-made for it all, let me tell you." She paused, took a sip of tea, then set her glass down. "You know what we could do? We could take Elena up there, and that dream healer—what's her name? Ginny. Ginny Woods." She reached down and pulled her tablet out of her enormous handbag. "I'm zapping my producer right this minute. And listen, honey," she said to Alice. "You'll come, too."

Alice pointed at herself. *"Moi? But why?"*

Maya gestured with her chin toward the neat rows being tilled by the gardeners. "Because by the time we put this together it will be, at the earliest, December, and you'll have a cosmetics line to launch, won't you, honey?"

Alice smiled. Then she leaned across the gap between them and squeezed both Maya's hands. "You aren't the Queen of the Night for nothing," she said.

Maya returned the squeeze, then turned to her tablet and tapped an icon. A moment later, her producer's face appeared on her screen. "Leslie? Listen, honey. Have I got an idea!"

TWENTY

ERROL SAT BEHIND his wide desk at the Veregro headquarters in Omaha as if he'd never left. He daydreamed about Alice. Being with her, especially now that she was her old self again, was as good as—no, better than—a mistress in every port. Alice was enough to make Errol monogamous. Or nearly monogamous.

That garden of hers had initially thrown him, but he'd quickly seen it wasn't a battle worth fighting. Not just Alice, but Maya, was invested in the idea. It would keep both of them out of trouble, which was definitely a good thing.

The tablet on his desk lit up. It was an anonymous text message, but Errol knew who it was from once he read it: "Winters has left Washington."

Errol allowed himself a thin smile. Bob Howard must have had Fisher show her the eminent domain petition. Errol initially had nixed the plan as risky, but her Achilles' heel had been as obvious as a birthmark. Still, Errol didn't like to admit that Howard had been right. After all, finding a way to take that damned Orchard away from her at all had been Errol's idea.

A knock on his open door grabbed his attention. Bob Howard, all 6' 3" of him, leaned easily against the jamb, arms crossed. "Heard you were back in town," he said, ambling in without waiting for an invitation. Cocky, that's what the guy was. Once this whole thing with Conroy was over, Errol was going to have some quiet talks with the other board members about replacing the CEO.

Howard sprawled into one of the visitors' chairs across from Errol. "What brings you back?" he asked. "You didn't like my eminent domain gambit?"

Errol wasn't going to give Howard credit that easily. "It was a risk," he told him. "Coming out in the open like that. She could have called our bluff, sicced the media on us, even called the authorities. I'd rather have handled it myself, on my own schedule. Now it's too late for that."

Howard offered an offhand shrug. "I'd hardly call it 'too late.' It worked, didn't it? She's heading home, tail between her legs. She won't be any trouble from here on out. I can guarantee it."

Errol leaned forward. "I'd be real careful about tossing around 'guarantees' when it comes to Charlie Winters, Howard. The woman is a loose cannon. You never know where her scattershot's going to land next."

"I'm not worried," Howard said. "Fisher says she lost her bark quicker than a loser in a dogfight."

Lord, in his next life, Errol hoped to be spared not just Bob Howard, but his homey witticisms. "Well, then, thanks," he said. "Is there anything else?"

Howard stood, shoved his hands in pockets. "How long you here for?" he asked, casually.

Errol shrugged. "As long as it takes," he said.

"As long as what takes?"

Errol smiled. "All of it."

Simon and his team, Samarie, Rafi, Todd and J.P., sat along the long table, each typing on a tablet. They were busy creating pamphlets and handouts for the Orchard's many visitors. The sixth seat sat empty. After Lottie had called to say she was coming home, Walter had left the rec room. Something in his face had suggested no one pursue him, and so no one had.

In a gesture that had become second nature, Samarie put one hand on her belly. Now that she was three months along, her mood swings had stabilized. Plus, she was hungry all the time. She dug into the bowl of dried fruit by her tablet and downed another handful.

Her tablet lit up. Maybe her mother was calling. Samarie swallowed quickly and tapped "answer" without even reading the display. O. M. G. It was Maya King. Samarie nearly choked on the dried fruit.

"Hi there, honey," Maya said. "I'm looking for Charlie Winters. I'm Maya King."

"I know you are," Samarie managed. "It's Lottie. Lottie Winters."

Maya looked surprised. "You're Charlie...Lottie Winters? Sorry, honey. I thought she was a lot... um...older. My bad." She laughed her familiar deep-throated chuckle. "That diet you're on must be even better than rumor has it."

Samarie shook her head. "No. I'm Samarie. Her daughter. It's just that she's Lottie. She doesn't like to be called Charlie." It was funny that Samarie didn't know why. She'd have to ask her mother about that.

Maya kept smiling. "Is she there?" she asked.

"Um, no. Not yet. We think she'll be back this afternoon. Can I take a message?"

"You sure can. Do you think, maybe, she'd be interested in having a show up there?" She consulted something on her tablet. "We're probably looking at early January. A new year's show, about the back-to-earth movement, organic gardening, self-healing, dream therapy, all of that."

"That would be so awesome!"

Another smile. "I hope your mother will feel the same way."

Samarie thought about it. Her mother probably wouldn't. Feel the same way. "I don't know..." she said.

"Well, then, maybe I've reached the right Winters after all. How good are you at talking your Mama into things?"

Samarie shook her head. "Not good at all. She's pretty...tough."

"That's good," Maya said. "I like tough women. I can see you're one, too. How about you think about how you can talk to her about it, and then have her call me? I'll bet between you and me, we can show her how important it is for women everywhere to be a part of a show like this."

"For women everywhere," Samarie echoed. "Okay."

"I'm sending you all my contact info now," Maya said, and a business card appeared in the corner of

Samarie's screen. "You call me whenever you want. We can make this happen, Samarie."

She'd remembered her name! Samarie smiled. "I believe you," she said. "I love what you do. We didn't even have access to media until a month ago, but I never miss your shows now."

"I'm so happy to hear that. Young women like you are our legacy, you know? Thank you. You'll call me, yes?"

Samarie nodded. They said goodbye. Then she sat, staring at her screen. Maya King. Holy shit. If her mother didn't go for this, Samarie would tie her to a chair until she saw how much it would mean to...to everyone.

Lottie had taken the Airporter as far as San Rafael, where Pete picked her up. When they arrived at the Orchard, she went into the rec room. Samarie, Rafi, Simon, Todd, and J.P. turned expectantly when they heard the door swing open. Samarie and J.P. leapt up and ran to hug her.

They all could tell how tired she was, just by her posture. Still, she stroked Samarie's hair, brushed J.P.'s out of his eyes.

"Maya King called!" Samarie told her.

"Who?" Lottie asked.

"God, Mom. You're kidding, right? She's only the Queen of the Night. The most influential talk show host ever."

"Can it wait?" Lottie asked, her voice weary.

"Okay, I'll tell you later. I'll bet you want to go lie down."

Lottie nodded. "Where's Walter?"

Samarie shrugged. "We think he's in his bungalow. He left the rec room this morning and we haven't seen him since."

"Thanks," Lottie left and made her way to the guest bungalows. She knocked on Walter's door.

Walter answered. "Welcome home," he smiled. "How did the mission go?"

"Becky stonewalled me and Patricia canceled our meeting." Lottie went inside and shut the door. "I think Errol got to them both before I showed up. He knew about the meetings, Walter. And he blackmailed them or bribed them, or persuaded them, or whatever the fuck the legal jargon is these days.

"And, ready for the bad news?" Lottie asked. "I had an unplanned meeting. A Veregro lawyer named Rick Fisher knew exactly where to find me."

"I know Fisher," Walter interrupted. "Cold fish."

"They've got a petition for eminent domain against the Orchard," she said.

"That's absurd. They can't do that! They—"

"Hold on a second, Walter. They haven't filed it yet. They've merely generated it, and threatened me that they'll file it if I go public with the report. Any of it."

"They can't take the Orchard," Walter said. "You own it free and clear."

"They could if it was for the greater public good."

"You have got to be kidding me."

Walter shook his head. "Nothing's above Errol."

She added: "I think the leak is here, Walter. We have a mole."

"Lottie, you're not saying, after all this, that you think I'm, somehow..."

"No Walter. I don't think it's you. I think it's Simon."

"I don't know. I've gotten to know Simon. He seems like a good guy."

"What do we really know about Simon? He's a reporter. For all we know he works for Errol. We never checked him out. I was so stupid. I knew I shouldn't have trusted him when he came here digging for dirt."

"Lottie, hold on. Let me look into it. I'll see what I can find out about his past. Let's not jump to conclusions."

"Walter, if it's not you it can't be anyone else but Simon. Everyone else here is like family to one of us."

"I'll check him out," Walter repeated.

Walter stood, began pacing the small room. "We just have to talk to the right people," he said. "Todd's been compiling a contact list of those who support what we're doing. We can start a grassroots movement, call their bluff..."

Lottie looked at him. Something crossed her features. Resignation, or loss. "I'm not sure we make such a great team, Walter."

"Lottie. Listen. This isn't over yet. We just have to keep plowing through. They won't take the Orchard. They can't."

"That's easy for you to say. This isn't your home. This isn't your family."

"No, I lost my family. But this is bigger than you, or me, or even Veregro. This is our world, Lottie. Our planet. It's too late to quit."

Walter was right. It wasn't about him or her. It was about Samarie and J.P., Rafi, her future grandbaby, all of the kids here at the Orchard. They deserved a future. And she was damned if she was going to let Veregro take that away from them. If she couldn't save the world, she sure as hell was going to save the Orchard.

"You need to go, Walter. If you stay here, Veregro targets the Orchard. You have to go."

TWENTY ONE

ERROL AND MAYA SAT at one end of the long conference table in Foster Media's Santa Barbara office, a compromise they'd come to years before so that neither had to sit across from the other's desk. Errol might be the boss, but as his chief moneymaker, Maya held more cards than the rest of his employees combined.

Like all their September meetings, this one concerned story ideas for Maya's talk show, brainstorming for special promotions, and a tossing back and forth of charitable foundations one or the other wanted to very publicly support that season.

As no cure had been found for IIS, they quickly agreed to continue to make finding a cure their main cause, at least through November, when they'd add their usual holiday charities. It was then that Errol insisted Maya host a special about Veregro's as-yet-unnamed world hunger initiative.

"You know how I feel about mixing your other businesses into my work," Maya reminded him, tapping a red nail on the tabletop for emphasis.

"And you know how I feel about signing your paycheck," Errol countered.

"I could leave and take my following with me," Maya said, softening her words with a smile.

Errol returned the smile. "Not according to your non-compete agreement you can't," he reminded her.

Maya laughed. "Touché, boss. Once your people have got a solid campaign in place, have them send it on to me. Since it's about food, let's shoot for Thanksgiving, shall we?"

Errol nodded. "Good idea." They both typed notes onto their tablets, then looked up. "What else you got cooking for fall?" Errol asked.

Maya consulted her tablet. "Well, there's Paris Fashion Week. You know I hate to miss that. And I want to do a show on organic cosmetics. I won't single out Alice—that would be pretty transparent—but we'll arrange it so the show will give her new line a definite leg up."

"Good," Errol said. "We need to look out for our own."

Maya looked at her tablet again, then back at Errol. "We'll also be doing a show about celebrities who have given up their careers to do good. Kelly Carmine in Mali, Stu Hall in Myanmar, that kind of thing."

Errol gave another nod.

"But listen. This one's my favorite. It's gonna be a whole week—probably in December, but maybe January, depending on how long it takes to get all the ducks in a row. A whole week on the back-to-the-earth movement. I've got Ginny Woods, the intuitive healer, on board, and Elena Medina, the organic gardener who organized the women's garden project in Ojai—"

"—the one who's responsible for my lawn being replaced by vegetables?" Errol interrupted to ask.

Maya laughed. "That's the one. But here's the best part. I've got a line on Charlie Winters, up at the Orchard."

"You spoke to Charlie?"

"Not yet. But I talked to her daughter—"

Errol looked up, sharply. "She has a daughter?"

Maya nodded. "Samarie. Smart cookie. She's expecting, due in December, I think."

Errol tried to cover his interest with a studied casualness. "I imagine her kid is pretty young," Errol fished.

"Just turned eighteen. Healthy as a horse. It's more proof of how well their diet works."

Errol sniffed. "Dumb hippies."

Maya shook her head. "They're not dumb, and they're not hippies, Errol. No one there's gotten MODs or IIS."

"Now you're quoting Walter Conroy to me."

"Sorry. It's so hard keeping up with your taboos of the week."

"Don't get smart, Maya."

It was Maya's turn to sniff.

Errol looked at his tablet, but wasn't really searching for anything. He was thinking about the girl, Samarie. He'd had a one-night stand with her mother eighteen years ago. Could the girl be a Foster? It was unlikely Charlie Winters would allow her to be DNA'd, but there were other ways of finding out if he was her father.

"Errol?"

"Hmm? Oh, Maya. Where were we?"

"My back-to-the-earth show. I want to go up to the Orchard for a retreat, film there. Samarie loves the idea, but of course, it's up to her mom."

"Charlie Winters used to work for me."

"I know her connection to Veregro, Errol. I'll make sure there's no conflict of interest. We'll be there to talk about the Orchard and women. Don't worry." Maya tossed her blonde curls back.

"I'm not worried, Maya. I know you'll keep it professional."

"Well so far I've yet to get a call returned."

"You're Maya King. She'll call."

Maya shut off her tablet, stood, and offered Errol a handshake (hugs were for home). "We all set, then?" she asked.

Errol nodded. "You keep me posted." He watched her make her way down the hall until she'd passed the conference room windows, then sat down again and swiped his tablet back on. It was time to give Charlie a call.

Walter flew back home to Omaha, rented a car and drove to his house. A real estate agent met him there and they discussed selling his home. He planned to use every last cent of any profit on the house to bring Veregro down. He wasn't going to fail. It wasn't an option.

"So is there any particular reason this needs to be a pocket listing?" the agent asked.

"I'm a very private person," Walter said. "I don't want my name or address showing up in the listings."

The agent looked around. "Well, I've got a few clients looking in this area. I'll start there."

Walter handed him the key. "Show it whenever you want. I won't be living here."

"All right," said the agent, pocketing the key. "You'll be taking the furniture with you?"

"I'm clearing out my personal belongings today, the rest you can sell with the house. Or give to charity. It makes no difference to me, but I'll pay you to take care of it."

"Fair enough. I think that's all for now," the agent smiled.

"Great." Walter reached out and they shook hands.

"Oh, one thing." the agent said. "I realize I only have an email address for you. Is there a phone number where I can reach you?"

"Sorry, no," Walter said. You email me and I'll email you back or call you within a few hours. Probably sooner."

"You are a very private person, Mr. Conroy."

"Very," Walter said, walking the agent to the door.

Walter was sure Kylie would understand his decision to sell their house and everything in it. She would support his desire to right what he'd wronged. Of course, if Kylie were still alive, he would have done nothing of the kind. If Kylie were still alive, they'd have their baby by now. Walter would still be working for Veregro, Randy's news one more message that he'd diplomatically listened to and then deleted.

Walter went to his liquor cabinet and found a bottle of single malt that was gathering dust. He'd been saving it for a special occasion. Today will do, he thought. He opened it, poured some scotch and went to the shelf where Kylie kept their photo albums. Not many people kept physical albums anymore, but it was something Kylie insisted on. She had remembered being a child and going through her parents' photos and wanted their children to have the same experience. Walter grabbed their wedding album and plopped onto the couch.

There they stood on a beach in Maui. Walter in shorts and a Hawaiian shirt. Kylie in a sarong and bikini. It had been Kylie's idea to ditch their large wedding plans.

Walter remembered they had been with their wedding planner picking out invitations.

"Are you committed to this font?" Angie, their wedding planner, asked them. "Because I want to show you something that's a little less Gothic, but still formal, and very pretty. I think you'll like it." She pulled out another sample.

"Um, I like it," Kylie said. "Walter?"

"Sure, very nice."

"And what about this one?" Angie asked.

"Walter?" Kylie said.

"I, I don't really see the difference between this one and the last one."

"Take a look at the serifs on this first one. They're so thin" Angie said pointing to the sample. "If we print thermographic we'll be fine, but if we go with litho, we'll never hold that detail. Since you guys really haven't decided yet, I'm thinking maybe this second one is best. Unless you're ready to commit to thermo."

Kylie and Walter looked at each other. "Try not to get overwhelmed, guys," Angie said. "This isn't even the hard part. Wait until you get to the seating assignments."

Kylie dropped the sample from her hand and stood up. "Why are we doing this, Walter?"

Walter stood up and grabbed Kylie's hand. "Because we're getting married. Isn't this what people do when they're planning a wedding?"

"I think so. But it just, kind of..." Kylie reached for the words.

"Sucks," Walter said.

"Exactly," Kylie said. "This sucks. Where's the fun?" Kylie clutched Walter's hands tighter and looked him in the eye. "Walter, do you take me to be your lawfully wedded wife, for better or for worse, from this day forward?"

"I do," Walter said.

"I do, too," Kylie said. "The rest of this is bullshit. What do you say we fly to Hawaii this weekend and make it official?"

"I'm all in," Walter laughed.

"Then kiss me." After, Kylie turned to the wedding planner. "Angie, we're still going to pay you. But you're fired."

Walter and Kylie flew to Hawaii and got married, barefoot on the beach. After the ceremony, Walter chased Kylie into the water. They splashed and laughed and ate and drank and had the time of their lives. That was Kylie. Walter closed the album, closed his eyes, and cried.

Until his tablet came to life. Incoming call. Ian Dunford. Reporter for The New York Times. Walter swigged the rest of his scotch and answered the call.

"Ian. You got anything for me?"

"This is all off the record, right, Walter? I don't wanna get hung for this."

"Haven't I fed you plenty of stories?" Walter asked.

"I'm not complaining, Walter. I just gotta watch my back. I had to cash in a big chip to get this info. An FBI chip."

"It's off the record, Dunford. Let's hear it," Walter said.

"Well, your hunch is right. Simon Keller smells funny. No fixed address, no phone number. All his mail goes to a post office box, has for years. He moves around a lot. Looks like he hasn't stayed in the same place for longer than six months. At least not in the last ten years."

"Where was he ten years ago?" Walter asked.

"Federal prison," Dunford said.

"For?"

"Dunno. Record sealed. Ten-year sentence, but he was out in three. This guy smells. You called it."

"Anything else?" Walter asked.

"That's all I got," Dunford said.

"Thanks, Dunford. I owe you one." Walter disconnected the call and tried to call Lottie but she didn't pick up. So he sent her a text message with the information he learned about Simon. He hoped the first three words would get her attention: Don't trust Simon.

Walter went upstairs to grab some clothing and stuffed them into a suitcase. Back downstairs he filled another bag with electronic gear: tablet, keyboard, wireless router, signal booster, chargers and battery back ups. Then photo albums, passport, title for the house. He looked around the living room and kitchen. What else? The rest of the single malt. Why not? He went to the kitchen and grabbed the bottle. The cap was missing, it was on the floor. Walter bent down to grab it and heard the shatter of breaking glass. Then more breaking glass. Hunched down in the kitchen, he heard a car peel away from in front of the house. It took him a moment to recognize he'd heard the sound of gunfire.

Walter crawled to the window next to the front door and peered outside. The street was empty. It was time to go. Walter loaded two suitcases and a duffel bag into the rental car, hopped in and cranked the engine. It would be less than four hours' drive to Salinas, Kansas, a good place to find a shitty motel, set up a wireless hot spot and start his blog. If he wasn't followed, he could stay a few days before he moved on. He wasn't sure how else to stay off Veregro's radar except to keep moving. Lottie was right. He was a target, and anyone around him was in danger.

Lottie's tablet was buzzing. It wasn't that long ago that she was free of this electronic burden but now that she had given in, she had a tablet that seemed to be constantly buzzing with calls and emails and text messages and private chats and incoming newsfeeds. An ever-expanding universe with no end. A digital black hole. How on earth did people deal with this day in and day out, she wondered? She reached for the tablet just to switch it off for a little while for some time to think, but the incoming caller gave her a shock. It was Errol Foster.

Lottie answered. "You have no case for eminent domain, Errol. My lawyers are all over it. You're blowing smoke, so back off." Probably a good time to get a lawyer, Lottie thought. She could afford it, just never needed one before now. Well, not since Angell Farm.

"It's been a long time, Charlie," Errol said with a smile.

"It's Lottie. I don't respond to Charlie anymore."

"You'll always be Charlie to me," Errol said, raising his eyebrows.

"Say it again and I'll hang up," Lottie said.

"Okay, have it your way. Lottie. You're still as aggressive as I remember. I like that."

"What is this? A social call?" Lottie asked.

"I suppose it is," Errol said. "I'm calling to ask how Samarie is doing. Pretty name."

Lottie felt a chill crawl across her skin. "That's none of your goddamn business."

"Really? I have this nagging suspicion that one simple DNA test would make it my business. And Samarie's business."

"You have nothing to do with Samarie, Errol. You're not her father."

"That would be disappointing, but I'd rest easier with some proof."

"Well, get used to disappointment. There'll be no test."

"I have fifteen billion dollars that says otherwise."

"Samarie wouldn't want your money, Errol. That's not how she was raised."

"You're missing my point completely," Errol said. "I will find out if Samarie is my daughter and you have no power to stop it. I will get to her, or I will get to someone who will get to her. I'll pay off the people who pick up your fucking trash and find a strand of her hair. I don't care how long it takes and I don't care how much it costs. No one is untouchable. All information has a price. You should remember that, Lottie."

"You haven't fooled me, Errol. I know someone here's been feeding you information and I know who it is. That game is over."

"Nothing's over. We still have a petition of eminent domain to discuss."

"Fuck you, Errol."

"Now you're talking."

"This conversation is over. Don't call me again. If you file the petition, I'll see you in court."

"You have no idea how hot you look when you're mad, Charlie."

Lottie picked up her tablet and threw it across the room. It hit the wall, bounced to the floor and the screen shattered. Her face was turning red and she tried not to scream. Her past had come back to haunt her. Had she really thought she could hide from it forever? She knew Errol would get to Samarie somehow.

She tried to think clearly. Steps. Break it down into steps. First, get rid of Simon. Walter had confirmed he was the mole. Then talk to Samarie. Time was up. The lease on the lie had expired. One way or another, Samarie was going to find out the truth. Errol Foster was her father.

INSIDE THE GARDEN OF EDEN
A blog by Simon Keller

I've been living inside this Garden of Eden for about three months now and I have gained ten pounds. On the surface that sounds like a bad thing but I was about ten pounds underweight when I got here. So what's changed?

If I name the stuff that pops quickly to mind it goes like this: I eat more meat. The cattle here are grass fed instead of grain fed, no hormones and no antibiotics. If a cave man ever ate a cow, it would have been like one of the cows here at the Orchard. I eat nuts instead of chips or any other fried snacks. Lots of fruits and vegetables, mostly raw or fermented. Very few dairy products and not much bread either. Dark chocolate when I feel like some candy. My coffee intake has mostly been replaced by tea. The pomegranate white tea is my favorite. If I feel like a sweetener, I use honey instead of processed sugar. As part of the honey harvesting team, I am particularly proud of my jar of honey. There is not a microwave in sight and nothing I eat has any artificial preservatives, only those provided by nature. I am outdoors more than ever before and my mode of transportation on the Orchard is primarily my feet.

In the upcoming weeks I will elaborate on my dietary and lifestyle changes because I am only scratching the surface here. But I bring it up now because what I found remarkable is that after gaining ten pounds on my new diet, my cholesterol level went down. What I notice about the others here is a high energy level from kids to adults. Obesity does not exist here, but some of the boys are quite large and muscular.

Oddly, none of the kids seems to suffer from acne here though it is quite common outside of the Orchard. And as I was told by some of the community in Bolinas, and now have seen for myself, no one seems to get sick here. Which brings me to MODS.

The MODS and IIS epidemics are spreading at alarming speed across the globe. Birth rates worldwide are at historic lows. This time last year, MODS was being compared to the HIV/AIDS epidemic which began in the 1980s. But now the rate at which MODS is occurring has HIV looking like a slow moving tortoise. The comparison now is closer to the Black Plague of the 1300s.

The link between GMOs and MODS and IIS is getting stronger. But still, no scientist has come up with incontrovertible proof, no smoking gun. Several studies are in progress to examine why Peru has only a handful of cases of MODS or IIS, but it will likely be years of gathering data before any results are reported. The only other locations on the planet with rates as low as Peru are Tasmania and Madagascar.

The common thread here is that all three locations have had a ban on GMOs for over a decade. Most other nations on the planet have no GMO-free zones in place. The Centers for Disease Control sent a team of scientists into the Amazon to study some of the lost tribes and see if they have been affected by MODS.

But all contact has been lost with the team; they have seemingly disappeared.

Here in the United States, the FDA and the EPA continue to back the use of approved GMOs and pesticides. They commonly refer to a case study of the African continent, once at the center of a hunger crisis that began in the 1980s and lasted for more than thirty years. After GMOs were introduced to the farmers, their crops began to double and even triple within ten years. By 2010 over ninety percent of all African farmers were using GMOs. By 2020 the hunger crisis was considered a problem of the past.

It's a nice story to highlight the positive side of GMOs but what the FDA and EPA do not address, and what no one can explain, is why Africa has one of the highest rates of MODS and infertility in the world.

So, is this all about what we have been eating?

Twenty Two

It was business as usual in the Orchard rec room. Simon was working on his next edition of Inside the Garden of Eden; J.P. was finessing some new graphics for the upcoming post; Rafi and Todd were doing research to help bolster some of Simon's writing with hard facts about MODS and its rapid spread around the globe, hitting highly populated areas the hardest.

Lottie came in and saw the group working. The kids had learned the technology so quickly it was scary. "Hey guys. Where's Samarie?" Lottie asked.

"I think she went to get something to eat," Rafi offered. "If you don't find her in the dining hall, then she's probably peeing. She's either eating or peeing. That's the routine."

"Gotcha," Lottie responded with raised eyebrows. "Rafi, why don't you and Todd and J.P. take a lunch break. Give me and Simon a little time to catch up on some things."

"Sure thing, Ms. Winters," Rafi said. "J.P., lunch time, buddy. Don't forget to command-S."

J.P. saved his file and the boys got up and left. Lottie kicked away the door stops, letting the large double doors slowly close, shutting out the sunlight.

"Uh oh. Why do I feel like I've just been called to the principal's office?" Simon asked.

Lottie forced a smile. "There's no fooling you, Simon. You're very perceptive."

"It's part of the job description. So, what's wrong?"

"I want to hand the reins of the blog over to Samarie. Her generation, they're the future here at the Orchard. You've given us a great start, but I think it's time now we continued on our own. It just makes sense. Samarie's been watching, learning. She's ready."

"But I don't want to go. I don't feel like I'm finished yet," Simon said.

"Oh, I'm sure you don't, Simon. Let's just part amicably now and leave it at that. We'll share profits on the blog's revenue through the end of the year. I'll honor that. I'm a woman of my word."

"It's not about the money. That's not why I want to stay."

Lottie gave a stifled laugh. "I'm not so easily fooled either, Simon. Money makes the world go round. I know why you want to stay."

"Okay, look. I need money. We all need money, right? But..."

Lottie interrupted. "But, some of us hold this Orchard sacred. And others will sell out anyone or anything for some money. You're a reporter, Simon. You report things when it benefits you. I'm sure it's all justifiable in your head. But you need to move on now and go report on something else. And be gone by morning." Lottie turned to leave.

"But this is my home now," Simon looked around the room. "This is my home. Why are you doing this?"

Lottie's patience evaporated and she whirled around to face Simon. "I'm trying to be civilized here so stop bullshitting me. I know about your shady past and I should have never trusted you. Go tell Errol you've been exposed. It's over."

"You think I'm working for Errol? That's not true."

"I don't believe you," Lottie said.

Simon spun around in his chair, reached into his backpack and pulled out his iGlasses. He held them out to Lottie. "Please put these on."

"Is this some sort of trick?" Lottie asked.

"It's the truth, that's what it is."

Lottie reached for the iGlasses and put them on. A blue light flashed in the lenses for a few seconds and then Errol appeared.

"Simon Keller, this is Errol Foster. Do you know who I am?"

"Everyone knows who you are, sir," Lottie heard Simon say. She was watching a playback of the conversation between Errol and Simon.

"That sounds like an offer too good to refuse," Simon said after Errol laid out his deal.

"You're damn right, Simon. You're never going to get a deal like this from James Murdoch or anyone else. This is a once-in-a-lifetime offer and it expires after this call. So you're coming to work for Foster Media?"

"Thank you for your generosity, sir," Simon said. "But I can't take the offer."

"Don't get greedy, Simon," Errol said as he swigged the rest of his drink.

"I'm not asking for more, sir. I've worked for large companies before and it's just not my thing."

"It's the way of the world, Simon. You're making a huge mistake."

"Probably, sir. It won't be my first huge mistake."

"But it will be your biggest. Last chance, Simon. Tick-tock, tick-tock."

"My mind is made up, sir. I like it here. I think we're doing a good thing and I want to stay on good terms."

Errol shook his head. "Foolish. One last thing, Simon. You breathe a word of this conversation to anyone, and I will ruin you. You understand? You'll be reporting for the Mojave Desert Times. And if you really piss me off, you'll be six feet under the Mojave."

Then Errol's image disappeared. Lottie pulled the glasses off her face and handed them to Simon.

"Can we trust each other now?" Simon asked. "I'm in your hands."

"You're not the mole," Lottie said.

"I'm not. Everything I've told you is the truth. I love it here. I love what we're doing. What we're trying to do. This really does feel like home to me now."

"I'm sorry I doubted you, Simon. Someone is feeding Errol information and you seemed like the most likely candidate. The only candidate, actually."

"I get it," Simon said. "But it's not me."

Lottie took a seat next to him. "Simon, where is your home? What about your family and friends? You must have a life outside the Orchard."

"It's kind of complicated," Simon said.

"Simon, if I'm going to allow you to stay here, and be around my children and the others here at the Orchard, I need to know more about you. Why were you in prison?"

"Shit," Simon closed his eyes tight and pushed his hair out of his face. "I'm not a violent criminal, Lottie. I've never hurt anyone. Not physically anyway."

"Why were you in prison?" Lottie asked point blank.

"I was never really in prison," Simon said.

"Explain," Lottie commanded.

"When I was seventeen I hacked into a NASA computer and downloaded the software that controls Legacy. It's a space station. I was playing around and evidently caused a twenty-one-day shut down of the NASA computer systems."

"Holy shit," Lottie said. "That sounds serious." Simon raised his eyebrows.

"But you were a kid," Lottie added. "I didn't think they put seventeen-year-olds in jail for offenses like that."

"Hold on, there's more. After that I hacked into a Federal Reserve Bank in Virginia and transferred ten million dollars to my account. That's what they arrested me for. They didn't know I was the guy who stole the NASA software until they searched my computers. I really didn't know what I was doing. It was all like a video game to me; one that I was really good at. I had no plans for the money. I wasn't funding terrorists or anything like that. It was just fun to see the number with all the zeroes show up in my little savings account. My bank blew the whistle." Lottie just nodded without saying a word and Simon continued. "So it was decided that I was a threat to national security, and the attorney general got approval to try me as an adult."

"What did your parents do? Could they afford a good lawyer for you?" Lottie asked.

Simon cast his eyes down. "Foster parents. They waved goodbye as two FBI agents handcuffed me and put me in a car. I never saw them again."

"My God, Simon," Lottie said. "They didn't try to help you or at least support you through the trial?"

"Are you kidding? They were just pissed off that their computers were taken away. But they didn't mean anything to me. I only knew them for a year. I can't even remember how many other foster homes I was in before them. So the trial. That was justice at its finest. It was a closed-door hearing, you know, because I was a threat to national security. No jury. The judge told me I was facing a ten-year sentence and the evidence was sitting in my bank account. So there was no defense. The prosecutor said he had a creative solution," Simon threw up quote marks in the air.

"So they offered you a plea bargain?" Lottie asked.

"I guess you could call it that. My sentence would be commuted to three if I served them under the command of the CIA." Lottie's eyes widened. "So I did," Simon said.

"Simon, are you telling me that you're a CIA agent?"

"I'm not. I'm just a reporter. But for three years I was property of the agency. I showed them how easy it was to poke holes into our government's software. Then I showed them how to plug those holes. I taught them how to hack into other governments' software: Russia, China, North Korea, Mexico. Jesus, Mexico," Simon let out a laugh. "I could have made a fortune consulting for them. Anyhow, I gathered intel on whomever or wherever I was told. No questions. I wasn't in prison; I was in a government compound. I had no friends, no family, just the agents who guarded me. Then one day they blindfolded me, put me in a car and dropped me off at a shopping mall. They handed me an envelope with some cash, said I'd served my time, and drove away."

"What did you do?" Lottie asked.

"I spent months looking for the place where they held me. I thought if I found it, maybe they would take me back, let me keep working. I was lonely there," Simon nodded. "But at least it felt like home. Anyway, it was hard to find the place. I estimated the drive to the shopping center was about four hours. So I planned my search around that. I finally found the place, but there was no one there. The place was abandoned. Mission complete, I guess. I didn't know what to do with myself. I just kind of roamed around to different cities until the money ran out. Then I figured I should be some type of investigator, or a reporter. See if I could use my skills for something good. So that's my story."

Lottie reached out and put her hand on his shoulder. "You're home now, Simon."

Samarie rested on her bed, door closed, reading a childbirth book that Felicia had given her, when she heard a soft knock on her door.

"I'm up," Samarie said.

Lottie opened the door and went into Samarie's room. "How are you?" she asked.

"I'm a little bit tired, a little bit hungry and I need to pee." Samarie swung her legs off the bed and plodded off to the bathroom. "What's up?" Samarie called from the bathroom.

"We need to talk about something," Lottie said. The toilet flushed and Samarie came shuffling out of the bathroom and plopped herself back on the bed.

"Okay," Samarie said. Now six months along, Samarie positively glowed. She was carrying mostly in front, a sign, Felicia said, that the baby was a girl. Lottie merely hoped for a healthy child; girl or boy, it would need to be strong to face what the world had left to offer.

"I've always felt strongly that the Orchard was the best place to raise you and your brother. I want you to venture out and see what's out there in the world. But I can't help but try and protect you from it, too."

"What are you talking about, Mom? Are you dying or something?" Samarie said with a small laugh.

"I've kept something from you, Samarie. And right now it does make me feel a little bit like I'm dying. I'm afraid you're going to hate me, but you need to hear this from me. I've waited too long to tell you. I know who your father is."

"You mean you found out who the anonymous donor is?"

"It's not an anonymous donor, Samarie. I wish that were true. Your father was a one-night thing. He wasn't someone I loved. But it turned out to be a blessing. I'm so glad I got you from that night."

"Is it Walter? You guys used to work together. I wouldn't mind if Walter was my dad," Samarie said, trying to smile.

Lottie squinted her eyes trying to trap the tears, and shook her head. "It's not Walter. Your dad is Errol Foster."

Samarie's smile slowly melted and she shook her head.

Lottie nodded.

Samarie sat for a moment with her hands covering her face, pressing on her eyes.

"Errol Foster? How could you keep this from me? My whole life you've known and you lied to me," Samarie said as she uncovered her face.

"I was trying to do what's best. To protect you. That's all I can say, Samarie. I know that's impossible for you to process right now. That man tried to ruin me and I had no idea what he might try to do if he knew about you. I couldn't risk it."

Samarie burst into tears. "Well, you could have told me."

"I know," Lottie said. "And I wish I had. By the time you were old enough to understand such things, I put it out of my head. It's a terrible excuse, I know. But I'm trying to be honest with you now. I'm sorry, Samarie. I wish I could take it back."

"If he's such a horrible person who wants to ruin my life, why tell me now? Why tell me at all?" Samarie said, shaking her fists.

"Because he knows about you."

"What? How long has he known about me?" Samarie demanded.

"I don't know for sure. Not long, I think."

"Did you talk to him?" Samarie asked, starting to calm herself.

"Briefly."

"And?"

"And he wants a DNA test to know if he's your father. But I know it's him. It could only be him. I didn't tell him that, though. I wanted to talk to you first."

Samarie rocked slowly back and forth on her bed. "What am I supposed to do with this now?"

"You need to think on it, Samarie. Decide if you want to talk to him. Decide if you want him to know for sure. Or maybe just let life go on as it has. It's up to you. But I think he'll try to reach you and I didn't want you to be surprised."

"Oh, like I'm not fucking surprised now?"

"I hope someday you'll understand, Samarie," Lottie said. "Would you like me to leave you alone?"

"I want you to make this go away. Tell me this is a bad dream. Tell me that my dad is a nice guy or a war hero or someone else. Not Errol Foster, the biggest asshole on the planet."

"I would make it go away if I could, Samarie. I wish it as much as you do," Lottie reached out to embrace Samarie, but she pushed her mother away.

"What about J.P.?"

"What about J.P.?" Lottie echoed.

"Do you know his father, too? What's his big surprise?"

The phone rang but Lottie didn't flinch. "Please answer it. It's probably Rafi checking up on me. I want to see him," Samarie said.

Lottie got up and went into the other room to pick up the phone, relieved at her good fortune that it rang.

"Hello?" Lottie answered.

"Hi, is this Lottie Winters?" the voice on the other end asked.

"This is Lottie."

"This is Maya King. You're a tough woman to reach, Ms. Winters," Maya said with a little laugh. "Do you mind if I call you Lottie?"

"No, that's fine."

"Great, and call me Maya. I hope you got my message. I spoke with that terrific daughter of yours."

"She told me. I've been meaning to call you back. It's just been so busy around here."

"Believe me, Lottie. I understand. I am busy twenty-four seven. So what do you think about hosting a women's retreat to be featured on the show?"

"I've got to be honest, Ms. King..."

"Please, call me Maya," she interrupted.

Samarie's ears perked up when she heard her mother say Ms. King. She got up and looked out of her bedroom to see her mom on the phone.

"Well Maya, the situation is..." Lottie paused when she saw Samarie standing at her bedroom door with a weak smile and tear-stained face. "The situation is that we'd love you to come here and do your show."

FORMER VEREGRO VP BIPOLAR, COMPANY REVEALS

Omaha. October 2, 2030. In response to questions regarding former Veregro VP Walter Conroy's allegations about the company's products, current VP of Communications Katrina Stark reluctantly revealed Conroy's history of bipolar disorder. "While it's Veregro's policy to never reveal a former employee's health history, Mr. Conroy's pattern of fabrications concerning the company have forced us to reveal his troubled past, including a reluctance to take his prescribed meds..."

Walter was packing and getting ready to move on to a new town when the article came up on his tablet. He punched the delete key, but the story lingered for a nanosecond before disappearing from his screen. How low were they willing to stoop? And Katrina— what were they paying her to diss her old boss? Had she no shame?

Walter knew the answers to all of these questions. Hadn't he made similar decisions when he'd sat in Katrina's chair? Who was he to judge her?

While his first reaction was to fire off a retraction and post it on his blog, he knew that was exactly what Foster and Howard were after. Better to ignore it and just keep on with what he was doing—funding testing at the only independent lab willing to align with him, then publishing the results as they became available.

His blog had begun to have a following. It helped, he knew, that he kept a sense of humor about it all.

So many crusaders preached from on high, so serious it was hard not to laugh at them. Walter had made the decision to keep his blog lighter despite its dire content, so they'd laugh with him, not at him. It had been a good call. As his mother had always said, you catch more flies with honey.

Another alert popped up in the corner of his screen, and Walter smiled when he saw it was from Lottie. He had written her a small note every day to check up on her and Samarie, check up on things at the Orchard. Lottie never responded except when he sent the news about Simon, she gave a curt "Thanks. I'll handle it." Walter read Lottie's note.

"Hi Walter. We're all doing fine here. Simon's still with us, not a mole, but Errol did try to get to him. Long story... Samarie's getting large and contributing to the blog. She says hi. The Orchard is going to be featured on the Maya King show. Not my idea but it's important to Samarie. Hope you are well. Stay safe. Lottie."

Before Walter could respond, a zap from Zach Kallen, the head of his lab, appeared. Walter clicked it open.

This looks definitive, read the subject line. *V + V. Will zap results once compiled.*

Walter hit "reply," then started typing. *Send what you have now. I can compile.*

After he hit "send," he waited, but after a minute with no response, he clicked Zach's tablet number. Zach answered, but he was looking away from his tablet while he spoke. "There's nothing to send you yet, Walter. I should have waited. I knew you'd jump the gun."

"Well, tell me, then," Walter said.

Zach continued to focus off to the side of his tablet. "You won't release this without the compilation?"

"Of course not," Walter lied.

"Okay. Those ten third gen mice we fed V+V? All ten are showing symptoms of MODS. All of them."

Walter nodded. "And the control group?"

"Nada. But Walter? There are still some deviations and variables I need to account for. Don't jump the gun, okay?"

Walter nodded again.

"Don't fucking lie to me, Walter. This is my rep we're talking about."

Zach was right. Walter needed to wait. "How long before you can get it to me?" he asked.

"Tomorrow night?" Zach suggested.

"You can do better."

"There are a lot of variables, Walter. You know that—you're the one who set up the protocols. I want to check and double-check. You'd do the same thing, if it was your lab."

Walter sighed. "Okay. But the second you've finished checking, you zap it all right to me, you got that?"

"I got it," Zach said. Now he did look at his screen. "Promise me you won't jump the gun," he added, yet again.

Walter held up his fingers in Scout's Honor. "I promise," he said. Maybe he even meant it. But time was of the essence. He'd at least leak a hint. That wouldn't compromise anything, now, would it? He'd give it some thought on his way to Grand Junction.

It was time to hit the road, but first he owed Lottie a response. He kept it brief.

"You were right to have me leave. Anyone around me is in danger. I'm on the move every few days. Hug to Samarie. Congrats on the Maya King show. So, who's the mole?"

TWENTY THREE

IN THE MONTHS since Walter had left, Lottie had come to rely on Simon and Rafi more and more. Simon set up email accounts, kept all the software updated, made sure their wireless connections were strong and reliable and still made time to focus on the blog. The blog grew fast and Samarie had become a regular contributor. Simon said they had several million followers. Some days, as she read through the responses, she believed him. Simon also followed Walter's blog closely and made sure Lottie stayed informed.

While the Orchard's blog rarely strayed from stories about Orchard lifestyle, Walter's was more direct, posting test results that corroborated, or even went further than, his and Randy's. Walter pointed his finger straight at Veregro, which until now had mostly ignored him.

Lottie still felt bad about letting Walter step into the crosshairs alone, but she knew her decision was the right one, the only one, for her. Except for her conversation with Errol, no one had so much as whispered *eminent domain* to her since she and Walter had gone their separate ways. And, while the Orchard's blog touted the virtues of naturally grown food, she never targeted Veregro. She wasn't going to sacrifice everything she'd built, or risk the safety of anyone at the Orchard.

Rafi was helping with the expansion of the Orchard. They needed more guest quarters; the kids wanted their rec room back, so it was time for proper office space and a meeting room. Lottie was insistent that these new structures fit into the current landscape, not the other way around. Trees would not be cut down, wires would not be strung from building to building, animals would not lose their space. The integrity of the Orchard must remain. Rafi worked closely with Pete to make sure Lottie's standards were met.

It was hard, but Lottie refrained from asking Samarie if she had made any decision about Errol. Neither of them mentioned it but Samarie was cold and distant when Lottie was around. The wound was still open and deep. Had she even told Rafi, or J.P. or Bren, or was she holding it all inside, Lottie wondered?

She tried to focus on writing the blog, on responding to comments, on planning for Maya King's visit in less than two months. But seeing Samarie always undid her resolve. It was her past coming back to tangle with her present. The breach between her and Samarie was her own doing, just as the Angell's deaths had been. Lottie missed her. But she deserved this. And she'd always known it was coming.

In the corner office reserved for Veregro's CEO, Errol Foster surveyed his domain. Golden parachute in his briefcase, Bob Howard had exited with little fuss, and, in his wake, the board had quickly elected Errol interim CEO until the fiscal year ended the following June.

From here, Errol could practically hear the hum of a successful, well-run empire, but it wasn't loud enough to drown out the insistent buzzing of that little gnat Walter Conroy.

Errol had never seen a good man go bad so fast. First, Conroy was running around with that crazy lab tech's test results as if they were the New Testament, then he'd hooked up with Charlie Winters and begun saturating the Web with misinformation. After he and Charlie split up—and Errol now knew from Samarie it had been the threat of eminent domain proceedings that had precipitated it—Charlie chose to share recipes and tree husbandry. But Walter went hog-wild, selling his house to finance a blog devoted to bringing Veregro down.

As if a gnat could bring down an empire. Still, a little proactivity never hurt.

Errol buzzed Marcia, whom he'd inherited along with Howard's office. "Get Katrina over here," he told her, then disconnected.

While he waited for the Communications VP to arrive in his office, he edited the press release she'd zapped to him that morning.

FOR IMMEDIATE RELEASE VEREGRO DISCOVERS INFERTILITY CURE

Working overtime at the insistence of new CEO Errol Foster, Veregro scientists have discovered that a byproduct of eel blubber enhances the reproductive cycle in mice. After further testing in rats, gerbils, and dogs elicited encouraging results, the company has requested expedited FDA approval of their new infertility drug, Fertigro.
     {ADD QUOTE} Foster said.

Errol tapped his tablet's keyboard onto the screen.

"Safety and world health are Veregro's primary concerns," Foster reiterated, he typed. *Since taking over Veregro in October, Foster has made infertility and the disease the twin foci of the lab's work.*

"Mr. Foster?" Errol looked up to see Marcia and Katrina at his door. He stood and walked around the desk, guiding Katrina in by her elbow.

"Let's sit here at the conference table," he said, parking her at one end, then retrieving his tablet before sitting down cattycorner to her. He tapped the tablet on. "I've just been reviewing your press release," he said. "Very good work. I've added a quote as you suggested, and then a bit more. Here. See what you think." He turned the tablet to face her.

Katrina read his brief additions, then looked up with a smile. "I don't know why you need me, when you write such great copy yourself," she said.

"Because a CEO hasn't got time to write copy day-in, day-out," Errol told her, the little twit. "But as long as you're here, let's tweak this first paragraph a bit, too, shall we?"

Katrina sat attentively, hands above her keypad. Once a secretary, always a secretary, Errol thought.

"For starters, let's delete this about eel blubber. It will gross people out. And this, too, about the reproductive cycle of mice. And don't mention animal testing at all. People don't like to be reminded we couldn't make the advances they need without the help of animals."

Katrina looked up. "That doesn't leave much," she said, timidly.

Errol offered her his kindly uncle smile. "I know you can make it work," he told her. "Make the fixes, add another paragraph or two, run it by me one more time, and then get it out by the end of the day."

Katrina nodded, then stood. "I do have one question, Mr. Foster, if you don't mind."

Errol was still seated at the conference table. "What's that, honey?"

"Jim Baker at R&D says Fertigro isn't anywhere near ready for FDA approval, let alone mass production or distribution."

Now Errol did stand. "Katrina, Katrina, Katrina," he said. "Public relations isn't about whether products are ready. It's about *relating* to the *public*. And right now, the public needs to know Veregro is working hard to solve its problems. As we always are, aren't we?"

Katrina smiled, nervously. "Of course. Yes. I see. I'll have something on your desk by 4."

Errol walked her to the door. "Make it 3, Katrina. Thanks."

He closed the door behind her. Christ. Walter Conroy had never questioned corporate's requests for early press releases. Walter had always gone along with the game plan.

Until he hadn't, that is. Well, let's just see what his little blog network did with this news. Errol could hardly wait to find out.

Marcia tapped at his door, then cracked it open. "Rick Fisher's here for your meeting," she purred.

"Right on time," Errol said as Fisher came into the office. "I like that."

"What can I do for you today, Mr. Foster?" Fisher asked as he sat down.

Errol slid his tablet across the table to Rick. "Tell me, have you seen this blog before?"

Fisher glanced at the tablet. It was Walter's blog. "Absolutely, sir. Well familiar. As head of your legal department, I can assure you we are always tracking sites like this that use Veregro's name is a disparaging light. Propaganda and lies, if you ask me, sir."

"Exactly. Now tell me, who hosts this blog?"

"Well the author is anonymous, sir. But it seems clear to me that Walter Conroy is behind it."

"Of course it's Conroy," Errol laughed. "That's not my question. What Internet provider is allowing this libelous crap online? That's my question."

"Off hand I have no idea. But I imagine it would be easy to find out."

"Do it."

"Yes, sir," Fisher said as he stood up. "I'll get back to you as soon as I know."

"Sit," Errol barked. "Find the answer now. We're buying the company and finding out where the posts are coming from. Then we're taking that blog offline."

"That could take some time, Mr. Foster."

"Well, get comfortable. This meeting's not over until it's done."

Walter read the release through twice. He knew the drill. He knew it was likely Fertigro was nowhere near ready for FDA approval. The question was, was Veregro going to push it through without the proper testing?

His source at Veregro would no longer talk to him. His source at the FDA had blocked his tablet number. Even though he knew it was unlikely to reveal anything, Walter tried an online search, first for the drug name (nothing), then for some of the test protocols (less than nothing).

That left him where he always seemed to end up these days, cowboying alone, far out on the range. Walter rather liked this image of himself, even if he actually lived in a small apartment in West Omaha.

Walter opened a new file. He began by pasting in the release, reading it again once he had.

FOR IMMEDIATE RELEASE
VEREGRO DISCOVERS CURE FOR INFERTILITY

Veregro CEO Errol Foster announced today that the company has found a cure for the IIS that's been ravaging the U.S. Working day and night at Foster's insistence, lab technicians have uncovered a complex formula that combats the dreaded plague.

Speaking from Veregro's Omaha headquarters, Foster said, "Safety and world health are Veregro's primary concerns." Since taking over Veregro in October, Foster has made infertility and the disease the twin foci of the lab's work.

FDA approval is pending. Veregro hopes to begin marketing the drug to an eager public by the new year.

Hmmm. That didn't sound like Katrina. Walter would bet anything Errol Foster had a hand in this composition.

Errol Foster. Not just the new interim CEO of Veregro, but the man behind the campaign to end IIS. Imagine, being able to hand yourself your own cash prize. No. That wouldn't be the motivation for Foster. But what would?

Winning. Winning by any means. That was what Foster was about.

So how could Walter bring not just him, but Veregro, down? There had to be some way of getting to the man, some way of exposing the sleights of hand that had held him harmless from his shenanigans all these years.

Walter remembered how close he'd come to being a victim of one such sleight of hand on that snowy Wyoming highway. And then again in his own home. Poor Randy hadn't been so lucky.

Walter had long ago picked the bones of Randy's research clean. Nobody read those posts anymore. It wasn't that no one believed them. The problem was they weren't sexy, or gory, or—

Gory. Randy. Walter had never written about Randy himself. He'd always felt that doing so would dishonor Randy's memory, in some way he couldn't quite articulate. But maybe he was wrong about that. Randy had died being true to his own code of honor, which was what had led him to share his findings in the first place. Wouldn't Randy want Walter to call attention to the mysterious circumstances of his death, if it helped further their cause? (Walter always thought of it as "their cause," now.)

He began to type. *What's wrong with this press release? Well, for starters, it was written by Veregro's CEO rather than its PR department. Why would the CEO of a company do something so menial?*

Because he's Errol Foster.

And why would Errol Foster do it?

Because he's got something to hide.

Randy Hall didn't die of the disease. Randy died because he published findings Errol Foster didn't want you to know about.

Walter didn't realize it, but he was now wearing a demonic grin. When he finished typing the entry, he read it through one more time, then tapped the key to post it.

Error 404: Sorry that page does not exist, his screen flashed. Walter tapped again to try and post. *Error 404: Sorry that page does not exist.* Walter sent in a service ticket online to find out what was wrong with the web page. Thirty minutes later he got a response that said his account was locked. They needed to speak to him. It was time to go.

Walter packed his gear and thought about his next move. What was Plan C going to be? Plan A was Lottie. Walter understood why she sent him away. She wanted to protect her family; protect the

Orchard. He knew he would do the same if Kylie and their baby were alive. What he would give to be back in Omaha, changing diapers and reading bedtime stories. Anything. Lottie tried to help and Walter felt grateful. But the plan failed.

Plan B was the blog. Go public. Rally enough public outcry to pressure an investigation. Then just let the facts sell themselves. It was all there; the reports were solid now. Confirmed independently. Only a scientist in Errol's pocket would deny it. Plan B was good but now the site was down. Errol had to be behind that. The web host wanted him to call so they could trace the call and tell Errol where to find him. Walter shook his head and wondered if he had grown paranoid. Maybe, but Errol had long arms in this world. It was all plausible. Walter wasn't going to risk it. Plan B was dead.

Plan C. Plan C. Plan C. Maybe it was time to disappear and leave the battle to somebody else. And then do what when the money runs out? What would Errol Foster do?

Before he could think of an answer there was a knock at his motel door. Walter was wrong. They didn't need to trace his calls to find him.

TWENTY FOUR

WINTER AT THE ORCHARD was usually a time for reflection. The rains replenished the springs and greened the meadows, and the Orchies spent more time indoors. This year, though, Lottie had devoted much of her December to making certain the guest cottages were ready to receive Maya King and her entourage.

Maya had explained that her crew, from producers to cameramen to gaffers, all used trailers, both for their equipment and as homes away from home. That would place all of them in town, where they could hook up more directly to the grid. But when Maya said she'd bring her own trailer, too, Lottie wouldn't hear of it. "We have four new guest cabins," she told Maya. "They're nothing fancy, but if you truly want to experience the life we live here on the Orchard, you won't find a better way to do so."

Maya had been delighted. She would occupy one of the guest cottages, Ginny Woods, the world-renowned intuitive counselor, another. A third would be shared by the community gardener Elena Medina and a young woman from Peru who was surrogating for an unnamed couple. A woman who'd recently founded an organic cosmetics line would be in the last cabin. Maya hadn't named names, but Lottie wasn't worried. Once she'd connected with Maya, she knew she could trust her implicitly.

Lottie had always valued her solitude, but, as she'd gone from cabin to cabin, making sure all was in order, she could feel her excitement growing at the upcoming visit from a group of women who, each in her own way, was trying to do precisely what she had done. Plus, Maya had insisted on including Samarie as well. Once Lottie conveyed this news, Samarie hugged her mother and then proceeded as if there hadn't been nearly two months of icy silence between them.

After checking the guest cabins, Lottie went to the rec room to see how Simon and his crew were doing. Simon, Rafi, J.P. and Todd were all hard at work. Noses practically pressed into their screens.

"Hey, gang. Take a break and try some of these pomelos I just picked. The winter crop this year is unbelievable," Lottie said as she put down a basket of coconut-sized yellow fruits.

Simon spun around in his chair. "Have you heard from Walter lately?"

Lottie cocked her head to the side. "No. Why?" she asked.

"His blog is offline."

"Maybe it's just a temporary outage or something," Lottie suggested.

"I don't think so. Other websites share the same host. I can reach them all. The IP isn't down, it's just his site. I think it's been taken down deliberately."

Lottie started to chew on her thumbnail.

"What's wrong, Lottie?" Simon asked.

"I'm worried something's happened to Walter. Not just his blog. You know Errol tried to kill him when he was on his way to the Orchard, right? I shouldn't have sent him away. If something happens to him now, it's my fault."

"Nothing's your fault, Lottie. You did what you had to do. Walter's a capable guy. Let's not assume the worst. I'll text him and let him know we're worried about him."

"I'll do the same," Lottie said.

"Walter's fine. We'll hear from him."

The first car up the long rutted road was a sleek silver Bentley. J.P., at the gate, saluted as he pointed the way toward the dining hall, where Lottie, Samarie, Rafi, Bren, and Simon had gathered to greet Maya and their other guests. Behind the Bentley bumped a satellite truck. Its driver offered J.P. a grimace; by this time, he was probably bone-tired of bumps and mud. An SUV brought up the rear. J.P. swung the gate closed and trotted up the hill in its wake.

Under a wide umbrella, Lottie stepped forward as a chauffeur opened the back door of the Bentley. The first to emerge was a very pregnant young woman with lustrous dark hair. Lottie instinctively offered her arm. "Welcome to the Orchard," she said. "I'm Lottie Winters."

"I am Sylvia Chavez," the young woman answered, lowering her eyes deferentially. Rafi, carrying another of the big umbrellas, moved forward to escort her into the dining hall, where Bren and her mother, Felicia, had set up a welcoming buffet harvested from the Orchard.

Ordinarily in rainy season, the Orchies pulled on mud boots before venturing outside, but with the influx of guests, Pete and Shooey had debated the best way to keep the paths relatively mud-free instead. Pete was ready to line up rough-sawn wood along every path; Shooey had rallied for creek gravel. It was Lottie who'd decided that Shooey's solution would be more attractive as well as more effective. As she watched Rafi escort Sylvia up the path, she was pleased to see she'd been right.

The next woman to emerge from the Bentley was as tall as Lottie, although probably fifteen years younger. The two assessed each other in the critical but friendly way women often do, Lottie noting the younger woman's immaculately styled honey-colored hair and wondering how badly frizzed her own auburn ponytail had become. Then she repeated her welcome, warmly.

When the woman smiled, Lottie saw how pretty she was. "I'm Alice Foster," the woman said, shaking Lottie's hand before impulsively embracing her. "Thank you so much for allowing so many of us to invade you."

Lottie barely heard the gracious thanks. Alice Foster? Line of cosmetics? Errol Foster's wife. Holy shit.

Maya King was next to emerge from the car, even smaller in person than she appeared on the screen. Charisma seemed to radiate from her in waves as she practically skipped to Lottie, then reached up and wrapped her in a warm hug. "Lottie! I can't tell you how delighted I am to meet you at last!" She turned as two more women were helped from the car. "This is Ginny Woods," she said as a stately woman with lovely long white hair joined them.

Lottie held Ginny's warm hand between her own. "The intuitive counselor," she said. "I'm honored."

"As am I," Ginny assured her. Simon stepped forward to escort her into the dining hall.

"And this is Elena Medina," Maya went on, "who manages a community garden in Ojai."

Lottie took Elena's hands and squeezed them. "We've been writing to each other for so long," she said. "It's such a pleasure to finally meet you in person."

Maya and Elena were escorted by Rafi and Simon into the dining hall. Meanwhile, rain-slickered crewmembers had come out of the satellite truck and SUV and, under the direction of a pretty young African-American woman, were busily hauling equipment in as well.

Samarie stood just under the eave, watching all the activity, a hand on her nine-months-pregnant belly. After he'd escorted Maya inside, Rafi came back out to join her. Standing side-by-side, they wrapped their arms around each other in a pose that had long become second nature.

"Oh, did you feel that?" Samarie asked, pressing Rafi's hand tighter to her belly.

Rafi laughed. "I felt it. That was a ninja's kick." He rubbed her tummy and laughed.

"Can't wait to meet ya," Samarie said, glancing down.

Lottie awoke after a sleepless night. One more day of preparation left and then they would begin the shoot for Maya's show. She and the other women of the Orchard had spent the evening getting acquainted with Maya and her group. Plenty of laughter and plenty of wine, but Lottie couldn't loosen up. She tried to stay in the conversations but she was lost inside her own head, wondering if they all knew her secrets. No. Why would they? The way the media had spun it, her history was with Veregro, not Errol Foster. He certainly wouldn't have told Maya, an employee, despite her own wealth and influence. It was unlikely he would have told his wife about his and Lottie's one-night fling. Her brief exchange with Alice would have had a different tenor if he had. Or perhaps Alice was just playing her. After all, she had no idea who Alice really was except the woman who married Errol Foster. Who would marry that man? Someone without a soul, that's who. Alice was not to be trusted, Lottie decided.

"Lottie, are you awake?" She heard Rafi call from outside before he pounded on the door.

"I'm up, Rafi. Door's open." She swung herself out of bed, threw on some sweatpants and a sweatshirt and greeted Rafi in the kitchen. She welcomed the distraction.

Rafi dangled a set of keys and wore a huge grin. "We're all set. Can I go get Simon?"

"Go get him. I'll meet you there," Lottie said.

Lottie went outside and walked the path from her house to one of the structures that had recently been completed. She carried a small stepladder and a brown paper bag. Everyone at the Orchard knew that new guesthouses had been constructed but this last building was a mystery. If anyone asked, Lottie just said, "You'll see."

Lottie finished screwing a second eyehook into the porch's overhang when Rafi and Simon walked up.

"You're just in time, guys," Lottie said. "Simon, there's a sign in that bag. Would you grab it for me?"

"Sure," Simon said as he stepped up the porch and reached into the bag. "Is that why Rafi brought me here? To help you with this sign?"

"Yep," Lottie said with her hand stretched out.

Simon was handing her the wooden sign when he noticed what it said: The Simon Keller Group. His jaw dropped.

"Give it here," Lottie said with a big grin.

"What is this?" Simon asked, handing the sign over.

Lottie fit the sign onto the eyehooks and stepped down from the ladder, dusting off her hands. "It's your new office, Simon."

Rafi dangled the keys and Simon grabbed them. He unlocked the door and all three of them stepped inside. It was a long, wide room. The floor and walls were made from natural bamboo, harvested from the Orchard's bamboo forest. Countertops made of recycled glass surrounded the room. There were cubbies and drawers and shelves. Plenty of space and connections to elegantly conceal the wiring, mount the screens and store books and files. In the center of the room was a large conference table.

"Do you like it?" Lottie asked. Simon looked around speechless. "Rafi designed it. So if you hate it, it's his fault."

Simon reached for Lottie and hugged her. Then he turned to Rafi and hugged him, too.

"I don't know what to say. I can't believe you did this. For me." Simon walked around the room, slid his hand across the cool, smooth countertops.

"You could say you love it," Lottie said.

"I love it," Simon said. "Wow. Just, wow."

"You deserve it, Simon. You've done a lot for us here and you've become a part of us. We want you to stay. And I'm glad to hear you love it. 'Cause it's moving day," Lottie nodded with a smile. "Alert your group. And if you have any trouble..." She nodded toward Rafi.

Rafi shook his head. He and Lottie said in unison, "It's Rafi's fault."

They all laughed and Lottie reached out to hug Rafi. He froze stiff like a surfboard for a moment, then relaxed and hugged her back.

"Thank you, Rafi. For so much."

Simon clasped his hands together. "Well, I think we'd better get busy. Rafi, would you mind organizing J.P. and Todd and start getting the rec room shut down? I'll be there shortly. I'm just going to look around here a bit more."

"See you there," Rafi said with an informal salute as he headed out the door.

As soon as Rafi was gone Simon grew serious. "Have you heard from Walter?" he asked Lottie.

"I haven't. I hoped you had," Lottie said, but Simon was already shaking his head.

"I'm worried now, too," Simon said. "The site's still down. I can tell that he didn't even open my text messages. I used a return receipt. They're unread. Or he's forwarding them to a new address so we don't know he's read them. But why would he try to hide that from us?"

"He wouldn't," Lottie said. "Something's wrong. Walter should have gone to the authorities for some protection or something. We were both so stupid. Errol isn't just some evil corporate overlord, he's a fucking murderer. Fisher was right. We're playing with fire. An inferno."

"Let's concentrate on that then."

"What do you mean?" Lottie asked.

"Foster's a murderer," Simon said. "We've been trying to fight Veregro, but Foster's an easier target."

Lottie froze as she took in Simon's assessment. He was right. They weren't playing with Errol on his own playing field. Words and reports were their weapons, not Errol's. Lottie nodded.

"Tell me who Foster's killed and I'll, you know, look into it," Simon said.

"We use our powers only for good, right?" Lottie asked.

"Only for good. Promise," Simon said as if he'd just pleased his mother.

"They got Walter, didn't they?" Lottie wasn't asking, just acknowledging.

"Maybe we'll hear from him any moment. I really hope so. But if Walter's still alive, Errol's not done with him yet."

After the Simon Keller Group relocated their gear to their new office, the rec room was transformed into the set for "The Maya King Show." Samarie arrived the first day of shooting and peered inside. The tables and benches had been pushed to the sides. In the center of the room, the women reclined on cushions. Samarie counted her mother, Maya King, Alice Foster, Sylvia Chavez, Elena Medina, Ginny Woods, Bren and her mother Amanda, the doula Felicia, the producer and director Leslie Irving, and three camerawomen. There were no men.

Samarie toed off her mudboots and hung her coat on a peg, then went to join the group. Before she could sit, though, her mother stood and touched her arm. "I need to talk to you," she said quietly.

Alarm coursed through Samarie's veins. "What? What's wrong?"

"Not here," Lottie whispered. "Excuse us for a moment, will you?" she said to Ginny, who nodded.

Lottie led Samarie to the kitchen at the back of the dining hall, then turned and leaned on a counter and crossed her arms.

"What?" Samarie said, leaning opposite her mother and crossing her own arms.

"You know that Alice is Alice Foster. Errol's wife," Lottie said. She grabbed Samarie's hand.

"Duh, Mom," Samarie said, sliding her hand away. "I know that's my stepmom." She smirked.

"Okay, let's not blow it out of proportion. She's Errol's wife. But you realize that baby..."

"Is my sibling. Or half-sibling, whatever. Of course I realize all this. Jeez."

257

"It's just that we haven't talked, Samarie. I wanted to give you your space so I haven't asked, but I think now is a good time to share notes."

"What do you want to know?" Samarie asked.

"Have you talked to Errol? Does Alice have any idea who you are?"

"Mom, I have a martial arts festival in my belly right now. If someone talks to me for more than thirty seconds, I have to excuse myself to go pee. I am overwhelmed with a capital O. I can't decide what I want to do about my brand-new dad. It's too much right now." Samarie started to cry. "I've hardly spoken to Alice or Maya or Sylvia or any of these people. I am trying to smile and nod and not completely fade out while they are talking to me. Can you understand that?"

"Sweetie, I understand perfectly. We're more alike than you will ever know. There's no rush. Not for any of this. I'm just grateful that you shared it with me, before we sit down with all these other women."

"So it's your turn to share, right?" Samarie asked, sniffling.

Lottie tried to hide her surprise. "Of course," she said, nodding her head just a moment after it instinctively shook.

"Walter's blog is down. Is he all right?"

"I don't know, Samarie. I really wish I did. I'm trying to reach him."

"Well, what could have happened to him?"

"Walter's situation is complicated."

"Errol tried to kill Walter, didn't he?"

Lottie took a breath, then chose not to speak. She just nodded.

"He was safe here. You never should have run him off," Samarie said.

Lottie looked down at the floor. "Walter understood why I asked him to leave. He wasn't safe here. Not anywhere. He'd only put us all in danger."

"What can we do?" Samarie asked.

Lottie paused before she spoke. "Sam? Let's just go out there and begin. They're waiting for us. We'll do what we can for Walter. Trust me on this, will you?"

Samarie dabbed at her eyes with her skirt. "Okay," she said. Then, arm in arm, she and Lottie returned to the dining hall.

TWENTY FIVE

LESLIE IRVING AND MAYA STOOD at the front of the rec hall and looked around the room. Everything was in place and filming was about to begin. "So what do you think?" Maya asked with a sweep of her arm.

"Well, it's not our most luxurious set, Maya, but it'll do. No go sit down and let's get this ball rolling." Maya trotted off and took her place with the other women on the cushions. Leslie clapped her hands a few times to quiet the room and gain everyone's attention. "All right everyone, take a last sip of water, clear your throat, and quiet down. We're ready to begin."

Leslie put a hand in the air, made eye contact with the camerawomen and each nodded. "5, 4, 3, 2, 1," Leslie counted down, then dropped her hand.

The three camerawomen, with their ear-mounted mini-cams, squatted quietly outside the circle as Maya began to speak. "We're here at the Orchard in Northern California because of the warm and gracious generosity of our host, Lottie Winters."

The camerawoman opposite Lottie turned to look at her, and a second later Lottie saw her face appear on the monitor mounted on the dining hall's wall. "We're honored to have you with us, Maya," she said. "All of you." She nodded to each of the other women in turn, and another camera recorded their faces, one at a time, as they returned the nod.

"Let's begin by introducing ourselves," Maya suggested. "I'm Maya King."

The others laughed. Everyone in the world knew who she was. Perhaps, everyone in the universe.

"To my right is the intuitive counselor Ginny Woods. I'll actually be passing the conductor's baton over to Ginny once we've finished our introductions, so it makes sense for you to meet her first. Many of you know Ginny from her books and lectures on the conscious feminine."

It was fascinating to watch Maya relate to the camera as if it were one viewer in her enormous audience. She was warm and familiar while at the same time commanding and self-assured. Now that she'd met the woman, Lottie could see why she had the following she did. If Lottie could muster even an ounce of what Maya exuded as a matter of course, Maya's audience would not only hear what Lottie had to tell them, but take it to heart—and home.

Ginny had finished her brief self-introduction and turned to Elena, who sat next to her. Elena had brought Lottie cuttings from some of the herbs she grew on the community farm in Ojai, and together, they'd set them out and said a little thank you to nature for them. The rest was up to the plants themselves. If the energy in this room were an indication, they were going to thrive.

Elena seemed shy of the camera until Ginny reached over and placed one of her hands over Elena's. The touch seemed to center and relax her, and when she turned to Alice, next to her, she reached for her hand as Ginny had taken her own. That set up a pattern the rest of the way around the circle: Alice to Sylvia, Sylvia to Samarie, Samarie to Lottie, Lottie to her friend (and Bren's mother) Amanda, Amanda to Bren, Bren to the doula Felicia, and Felicia back to Maya. By the time they had finished, they were all holding hands. Maya turned to Ginny and with a soft nod passed back the role of speaker to her.

"The energy of the feminine is inside each of us: man and woman, child and adult," Ginny began, head bowed, eyes closed. The others did the same without being told. "This week, we will turn inward so that we may turn outward. We will reach deep within ourselves so that we may share our wisdom and love with all of you. We thank Maya for sharing her time… and her audience."

The others laughed.

After acknowledging her little joke, Ginny shut her eyes again. "We thank Lottie for sharing her home, and her harvest. We thank Elena for sharing her wisdom, and her spice."

More laughter.

"We thank Alice for sharing her creative energy, and her vision. We thank Amanda and Bren for sharing their love of this place, and of each other. We thank Felicia for her gift of guiding new life into being, for her mother wisdom. And most of all, we thank Sylvia and Samarie, for sharing their bodies and their blossoming, with their soon-to-be-born babies, with all of us. We thank ourselves, we thank you, and we thank Gaia, for bringing us to this moment, and all the ones that follow."

"Thank you, Gaia," Maya echoed, and the others followed suit.

After a moment of silence, Ginny opened her eyes. "So," she said. "I've been thinking, ever since Maya asked me to orchestrate this retreat, how we might begin. But it wasn't until yesterday afternoon, when we arrived, that I saw the way I had perhaps always known. We are here, now, because this is where it all begins again. We are at this place at this moment because we *can* change the direction in which the world has been going. Around this circle, we have creativity, strength, love, nurturing, power—and money." Laughter. "Within this circle, we can create a path that leads us to ourselves and to our future."

The others nodded. Ginny turned to Lottie. "Lottie, why have you chosen this moment to share what you've nurtured with the world?"

Lottie was surprised it had come so soon. She shut her eyes tight. "Why have I chosen this moment?" She repeated it, opened her eyes and looked up. "The moment has chosen me. It all begins with my kids. Beautiful Samarie here with us; my son, J.P., whom many of you have met. I nurture a future generation. All of us here do, right? In one way or another. We nurture our kids and our pets and our plants and hopefully our food, our lifestyles, even our technology. Something I myself am just coming to understand."

The women laughed, and nodded in agreement.

"I've nurtured this Orchard," Lottie said. "My father taught me how. Charles Winters. He's the reason the Orchard is here. He saw the future. That we needed to return to nature. To nurture our nature; our mother earth. I always believed him but I took it for granted. But the spread of MODS has put it all in perspective. So I guess that's why my moment is now. Because of the fourth generation. I want to save them."

Maya was growing visibly nervous. Leslie looked confused. Lottie never talked about any fourth generation in their pre-interviews. They were supposed to be upbeat, positive and educational. Give people hope, for Christ's sake. Where was this going? Before she could interject and change the subject, Ginny spoke up.

"Lottie, for those of us who don't know, what do you mean about the fourth generation? What does that have to do with MODS?"

"I learned about it from a man named Walter Conroy. Until last April, Walter was the vice president of communications for Veregro. That's right—Veregro. As you probably know, I used to work for Veregro, too. I'd thought, back then, that Veregro could save the world. It appears Veregro still thinks that. Then Walter's wife died of MODS. She was only a few weeks away from delivering their first child. But the child died with her."

The women shut their eyes, to honor the loss of this sister.

Lottie waited a moment, then went on. "Walter was understandably distraught. But it was at this time that a Veregro lab tech approached him with some disturbing test results. The tech—his name was Randy Hall—had run these particular tests because he'd come across some that Walter had run years before, and was intrigued by what Walter's had shown. You see, Walter comes from a lab background. At heart he's a scientist.

"The tests—both Walter's and Randy's—showed, beyond the shadow of a doubt, that lowered birth rates, infertility, rashes—and every single one of the symptoms of MODS—could be traced to treated genetically modified organisms."

Sylvia let out a loud gasp. "What does this mean," she asked, "genetically modified organisms?"

It was Alice who answered her. "It means everything that Veregro's been feeding us for the past forty years." Her voice was even, but Lottie could feel the surprise and anger beneath her words.

Maya spoke up. "But forty years. That's a long time. How could they be so sure it was the genetically modified organisms?"

Lottie nodded. "That's why it's been so hard to prove. The treated GMOs didn't affect the first generation that ate them. Even in the second generation, symptoms could be attributed to a host of other syndromes, to autoimmune disorders, to genetic disease. But in the third generation, the symptoms were specific, uniform, and incontestable. Infertility, including the inability to carry to term and stillbirths; sterility, including dead or defective sperm. Bizarre skin problems. Specific organ failures, particularly the liver and kidneys. And death."

After a moment of stunned silence, Sylvia turned to Samarie. "So, Samarie...and myself? We are this third generation?"

"That's right," Lottie said. "But Samarie has never eaten anything but the Orchard diet, and I was raised here, too. It was my father who developed what we eat—and don't—in the first place, so, except for the years I was at college, and, briefly, in the workplace, I've never eaten anything else. And you, Sylvia, you're from Peru, which banned GMOs years ago. That's why each of you has been able to carry a baby to term. We're lucky to have two healthy women here carrying the fourth generation."

Sylvia began to cry. "But I am here for the whole time my baby is growing inside me. I am eating the food of America all this time." She turned to Alice. "If you know these things, why do you not stop me? This is your baby, too! Why do you not say? What are you thinking, Alice?"

Alice, too, began to cry. "I didn't know! Of course I didn't know! Do you think I would have risked my baby—risked you, Sylvia?—if I knew?" She stared off into middle distance for a moment before focusing in again. "It's my bastard of a husband," she said, her voice gone terrifying quiet. "Errol knows. Errol has always known."

"Okay, stop. Stop filming," Leslie shouted to the crew. The camerawomen touched their ear-cams and their small green lights turned red. "What the hell are we talking about here? None of this is in my notes." Leslie shook a stack of papers rolled up like a club.

Before anyone could stop her, Alice pushed herself up from her cushion and strode out of the dining hall. Through the window, the stunned women watched her run down the steps and into the rain. She looked right and then left before choosing a direction and disappearing from view.

Maya, who'd watched Alice bolt, turned back toward the circle. "Let's take a break, shall we?"

"That went well," Leslie said to no on in particular.

Elena reached across the space Alice had left to take one of Sylvia's hands. Samarie held the other. "Mr. Errol is not a good man," Sylvia said. "I always know this, but I tell myself I am wrong."

"It's not for us to judge others," Ginny said.

"I am not being to judge! I am speaking what is true! Oh!"

Felicia, who'd been sitting quietly next to Lottie, sprang across the circle in two steps, then squatted in front of Sylvia.

"Breathe deeply," she said, laying a warm hand on Sylvia's belly.

"Oh! The baby!"

Felicia lowered her hand, then exchanged a look with Ginny, who rose. "Sylvia's baby is coming," Ginny said. "Let us offer her our strength and love, our hope and energy."

"Oh!" Sylvia cried. "It is hurting me!"

Samarie turned away from Lottie so that she could lean toward Sylvia. "I'll stay with you," she said. "Don't be afraid."

Felicia stood and took Sylvia's hand. "Try to get up. I'll help you. We're going to the birthing house."

The other women rose and Lottie turned to Maya. "I'm sorry I ruined your show. I didn't mean to, but it's part of the story. It's part of the Orchard's history now. I don't blame you if you never air what I said, but you can help a lot of people if you do. You can wake the world up."

Maya looked deep into Lottie's eyes, and Lottie saw that she knew all about Errol. At the same time, she saw that Maya had decided that the opportunity to give what she could to the world was worth the bargain she'd made with that particular devil. But now she was questioning herself. She probably did that far more often than anyone ever imagined.

Lottie placed a hand on Maya's arm. "What you do is worth whatever compromises you've had to make to do it," she said.

Maya nodded. "I know that," she said. "But sometimes—times like this—it's hard." She shook her head, as if to clear it. "I need to call Errol and tell him the baby is coming. Can you find Alice?"

"I'll find her," Lottie offered a weak smile. "See you back at the birthing house."

Lottie searched the Orchard grounds for nearly an hour but never found Alice. She went by the birthing house to see if perhaps Alice showed up there. She found Maya on the porch, under the overhang to stay out of the rain and frantically typing into her tablet, almost as if she was playing some shoot-em-up video game. Maya looked when she saw Lottie approaching.

"Did you find Alice?" Maya asked.

"No. I hoped maybe she came here," Lottie said.

Maya shook her head. "No one's seen her. I've been texting her every five minutes. I reached Errol. He's on his way. Should be here in less than two hours. I couldn't tell him not to come, Lottie. It's his baby."

"I understand, Maya. Does Samarie know?"

"No, I haven't even told Sylvia yet. I was just about to head inside and tell her."

Lottie shook the rain off her coat, then she and Maya went inside the birthing house. Felicia and Samarie were sitting in a darkened room and Sylvia could be heard breathing in the adjacent room.

"Everything okay?" Lottie asked.

"Sylvia's dilating very slowly. We're just giving her some space right now to relax and try to get comfortable."

"We need to tell her that Errol is on his way. He'll be here in just a couple of hours."

Samarie, eyes wide, looked up sharply at Lottie, and Lottie gave a slow nod.

"What about Alice?" Felicia asked.

"We haven't found her yet, but we will. I'm sure she just needed some time by herself," Lottie said to Felicia. Then she turned to Samarie. "Should we talk?"

Samarie took a deep breath and shook her head.

"Okay. Be strong. I'm going back out there to find Alice."

Lottie pulled on the hoodie from her jacket and drew the strings in tight. The chill was getting to her after spending so much time outside in the rain. She left the birthing house and jogged over to Simon's office.

Simon looked over from his screen when he heard the door open. "Oh, thank God it's you. Close the door."

Lottie shut the door and took a seat next to Simon at one of the computer terminals. "Did you hear from Walter?" Lottie asked.

"I didn't. None of my emails or texts have been opened, still. But I have news on our little project."

"You found dirt on Errol."

"Better than dirt. Blood. I've connected him to the Randy Hall killing. We got him."

"Good. He'll be here in ninety minutes."

"What?" Simon nearly spilled out of his chair.

"That's right. His surrogate, Sylvia, went into labor a little while ago. Maya called Errol and he's on his way. Get ready to put on a show."

"I didn't think we were gonna confront him, like, personally. I thought we'd just leak this to the press or something."

"Well, we got lucky, Simon. Maybe we can find a shred of humanity left in him once we threaten to expose him to the world."

"What if he's armed or something. The guy might try to kill us."

"Don't worry about that. Pete will check him out. Errol's not getting in here with any weapons. But there's something I need to share with you because it's probably going to come out."

"It is about J.P.?" Simon asked.

"No," Lottie said. "It's about Samarie. Errol is her father."

"Holy shit."

"Tell me about it," Lottie said.

"Samarie knows?" Simon asked.

"She does. She's never met him though. It's gonna get weird around here."

"It's already weird."

"Simon, why did you ask about J.P.?"

"I thought maybe you were gonna tell me...," Simon shook his head. "Never mind, I shouldn't have opened my mouth."

"Tell me what's in that head of yours, Simon. We don't have time for bullshit right now."

"You know I'm on your side, Lottie. I would never tell anyone."

Lottie slammed her fist down on the counter and the keyboards shook. "Tell me what you know."

"I know that J.P. is John Paul Angell. The boy that survived the Angell Farm killings."

Twenty Six

SAMARIE WAS OUTSIDE the birthing house grabbing a breath of fresh air when Rafi came along with a bowl of chopped fruit and vegetables.

"Thanks, Rafi. You're a lifesaver," Samarie said, taking the bowl.

"Anything else you need, just let me know. How's it going in there?"

"Sylvia's baby is in no hurry to meet us all."

Rafi laughed and tickled Samarie's tummy. "I bet this one's anxious to meet us." Rafi heard the mechanical roar first, but couldn't immediately identify the sound. "What's that?" he asked. "Do you hear it?"

"I think it's a plane," Samarie said, looking up into the low sky trying to spot it through the rain.

"That's weird. Planes never fly over the Orchard," Rafi said. "And this one sounds like it's flying pretty low."

"It's Alice Foster's husband. He's the father of Sylvia's baby. I found out a little while ago that he was on his way here."

A moment later, the jet flew over again, lower this time.

"He's looking for a place to land," Rafi said.

"He's going to land at the local airfield and the SUV will go pick him up," Samarie said. They watched as one of the rain-slickered assistants, along with Pete, hopped in the SUV, turned it around, and began to bounce back down the road. "Rafi?"

Rafi ran an affectionate hand through Samarie's hair. "Yeah, gorgeous?"

"Ms. Foster's husband in that plane, he's Errol Foster."

"THE Errol Foster?"

Samarie looked down. "Yeah, THE Errol Foster. There's something about him you don't know. I just found out myself. I don't know how I feel about it. But... it turns out, Errol Foster is my dad."

"Oh. Wow," Rafi said. "That's intense. You're not joking around, right?"

Samarie shook her head. "My mom used to work for him, at Veregro. And they got together, and, you know. So, here I am."

"Yeah, I know," Rafi said. "I know how it works."

"I've been thinking about if I want to get to know him and call him dad. Or if I hate him. Or if I just want to pretend I never even knew about him. Let life just be as it was. What's the difference now? So, I don't know how I feel. Maybe numb from shock, or maybe, I'm feeling nothing."

"That's a lot of pressure, Sam. What does your mom say about it?"

"She says that she supports whatever I decide. But I know what she really thinks. He's our enemy, even though he's my dad."

"Sounds like the family reunion from hell is about to happen," Rafi said.

"When I was researching the Angell Farm story, I read everything I could find," Simon explained. "It seemed strange that there was no follow-up on the baby that survived. Nothing anywhere. I was curious about him. Did he recover from his injuries, did he succumb to them? What happened to that kid? It really haunted me. So I looked into it. You know, my way. Official records said the boy was adopted by a woman named Margaret Sinclair—the nurse who cared for J.P. at the hospital where he was brought."

"I know what the records say, Simon," Lottie said, pacing around the room. "So what did Margaret tell you?"

"Margaret wouldn't talk to me, she's kept your secret. But it was easy to figure out that she didn't raise the boy. Her story was suspicious to begin with. A thirty-eight-year-old nurse who quit her job four months after she supposedly adopted a baby boy. Then bought a $2 million home and hasn't worked since. Where does a nurse without a job get that kind of money? Maybe she has rich parents. But she doesn't, I checked it out. Maybe a rich ex-husband. Nope. Lottery winner? Uh-uh. So I watched her for a while to verify my suspicion. No sign of a son. No calls, no emails, no text messages. No mention of him on her weekly call to her sister in Connecticut. Then when I came here and met J.P., all the answers kind of fell into place."

"When I walked into the Angell's house and saw that family," Lottie said to Simon, losing the fight to keep her tears inside. "You can't possibly imagine, Simon. To see those kids, the blood. Those lives all lost because of me. When I saw that baby breathe, I knew he was my responsibility. I didn't care what I had to do, or give up. I knew I'd spend the rest of my life at the Orchard to keep it a secret. And it was the right choice." Lottie broke down into sobs.

Simon embraced her. "You're a strong woman, Lottie. J.P. has no idea how lucky he is. I would do anything to have had a mother like you. No one ever fought for me. No one."

Lottie sniffled and cast her eyes up. "It makes me think about my dad. I could never have done it without him. There was no chance in hell the state was going to let me adopt the boy. Not Charlie Winters, the woman who drove John Angell to violent madness. My dad spent most of his fortune to help me get J.P. Margaret helped us, and a few others. A big payday for them all. I wish he were here to see his grandkids, and to see how his Orchard has grown. And to see me, his little Charlie."

Simon was silent. He knew Lottie needed to cry, that she was lost in the memories swirling in her mind. After a few minutes, she took a deep breath and squeezed Simon's shoulder. "No one knows about this, Simon."

"We've both shared some secrets, Lottie," Simon said. "And yours are safe with me. I promise you. When I first came to the Orchard and tried to talk to you, I wasn't trying to expose anything about your personal life. I had no clue John Angell Jr. might be here. I sure wanted to find him, and I'm glad I learned what became of him. But, that's not what I came here to ask you about. I came to ask if you thought John Angell really killed his family."

The question momentarily stunned Lottie. "What do you mean? Of course he did. I know he did. I found them," she said. Her eyebrows raised in recoil. *I found the note,* Lottie thought. She had no doubt that John Angell was responsible.

"Not all of the investigators working the case were convinced he did it," Simon said. "I had a chance to read through the case files, and I wasn't convinced either. Reads more like an unsolved mystery to me. The police chief thought it was a clear case of murder/ suicide though, so he halted the investigation after two weeks and made the ruling. But he just wanted everything to return to normal in Des Moines. Get the press out of his face."

"I won't ask how you got your hands on the case files, Simon. But if you read them, then you must know everything I had to say on the matter. They were quite thorough when they questioned me."

"Didn't you think it was odd that Angell did such a sloppy job though?"

"What do you mean?"

"For starters, why was John Jr. left alive?" Simon shook his head and looked down. "I know that's a disturbing question. But just think this through with me for a minute. Seems unlikely Angell wanted his family to suffer, so why didn't he make sure he finished the job?"

"I don't know. I never thought about it. The man was obviously crazed. He couldn't have been thinking straight," Lottie said.

"Maybe. But he seemed to be thinking straight when he drugged the whole family and got them into their beds. Doing a sloppy job in the final stage seems, off. Plus there was no suicide note. One investigator argued that there was no note because John Angell and his family were murdered, and the murderer or murderers were in a hurry to get out of there. That's why they didn't check to make sure everyone was actually dead. Someone tried to make it look like a murder suicide, and did pretty good. But they left a couple of loose ends."

"I'm no investigator, Simon, but that sounds a little farfetched to me."

"Well, that's just a theory. There's still a very compelling question. An object, something, was taken or moved from the kitchen table where John was shot. After he was shot."

Lottie's heart started to race. She remembered that folded up piece of paper on the table with her name on it. She had been compelled to read it, and once she had, she knew she had to take it.

Simon described the scene. "The bloodstain pattern analysis from the investigation showed a rectangular area on the table that was clean, but should have been splattered with blood, like the rest of the table. It meant something was covering that spot when John was shot. Something almost flat, not tall. Maybe paper, or a placemat. But too small to be a placemat. So some of the investigators thought that was a good enough reason to keep the investigation open. It added to the theory that someone else murdered the Angells, because someone was there to take something. It didn't fly though. The chief overruled it; said the idea was too far out there. He reasoned one of the local cops moved something and forgot about it, a napkin or a paper towel. Very careless, whoever moved it before the scene was documented. But it's happened before. The thing is, nothing collected into evidence fit the spot, so investigators never found out what it was."

"Simon, the only murderer was John Angell. There was a suicide note on that table. I took it." She stared at Simon, and tried not to look away. "It was addressed to me. It was personal. So I took it. It wasn't going to change anything. I'm sure that's hard to understand."

Simon's eyes widened. "Do you still have it?"

Lottie slowly nodded.

"Holy shit, Lottie. Bring me that note."

The SUV carrying Errol returned to the Orchard, rolled past the rec room and the dining hall and proceeded to the guest cabins. The PA, Pete and Errol all hopped out of the car.

"Where's my wife, Alice? And our surrogate, Sylvia?" Errol asked Pete.

"Alice went for a walk. I'm sure she'll be along soon."

"In this weather? That's insane," Errol said.

Pete shrugged and ignored the comment. "Sylvia's in our birthing hut. I hear she's doing just fine. These are the guest quarters," Pete said, gesturing to the cabin where they parked. "Your wife is staying here. Sylvia's in the next one over."

Errol interrupted. "Take me to the birthing hut. Now."

"Sorry, can't do that. My orders are to take you here first so you can drop off your stuff, clean up, use the head if you need to and then I'm taking you to see your host."

"What are you talking about, my host?" Errol said.

"Your host, Ms. Winters. She's expecting you and I'll be bringing you to her. Now go do what you need to do inside and I'll wait here."

"I don't need to do anything. Take me to Charlie and let's get this formality over with."

Pete tensed his eyebrows. "You'll address her as Ms. Winters, or Lottie, if she allows it. Is that understood?"

"Do you know who you're talking to?" Errol said more than asked. "Who the fuck do you think you are?"

"I know who you are. Outside these gates you're some rich and power- hungry guy. Inside these gates, you're just an unwelcome guest. And I'll tell you who the fuck I am. Here at the Orchard, I'm the law." Pete took a step toward Errol, dwarfing him in size. "We understand each other now?" Pete started to head down the path toward Simon's office. "This way."

"This is bullshit," Errol scoffed as he tightened the hood on his rain slicker and followed along.

They stepped on the porch of Simon's office and Pete knocked on the door. Lottie opened up and smiled.

"Thanks, Pete," Lottie said.

Pete gave a nod. "He's clean, I checked him out. I'll be right outside if you need me."

Errol was shaking off the rain. "Come in and sit down," Lottie said, shutting the door behind him.

Errol looked around the room and noticed the row of wall-mounted monitors and the clean set up they had for their technology center. "Hmm, not bad for a hippie commune. I'm also surprised to see you run this place like North Korea, with your military goon outside there. What is this, some sort of cult you're running here?"

Lottie gave a polite laugh. "Welcome to the Orchard, Errol. You can be assured that most guests receive a different reception. Let's just say, you're special." Lottie pulled out a chair from the conference table at the center of the room. "Have a seat." Lottie gestured to Simon, who was sitting in front of one of the monitors. "I believe you know Simon."

Simon spun around in his chair and gave a grin.

"Oh, Simon Keller. The boy wonder. It's really hard to believe you turned down my offer for this place," Errol said wrinkling his nose. "You really are dumber than you look."

"We have a little presentation for you, Errol," Lottie said.

"Great. Entertain me. Then I'd like to see my wife and Sylvia."

"I don't think you'll find it entertaining," Lottie said. "And maybe we can avoid some of the ugliness. We'll see how you answer my questions. So let's talk first."

Errol slapped his knee and started to laugh. "Really, this is laughable. I get that you are the head honcho here, Lottie. You must feel very powerful to have my attention for a few minutes. No problem, I'll grant you that. But you do realize this is still the United States of America, right? This Orchard is not some other country. You have no power over me. So get on with your little show. I'm anxious to get out of here, with Alice, Sylvia and my new baby. See if we can find a decent place to stay around here."

"I want to know where Walter is," Lottie said.

"I can honestly tell you I have no idea. Well, that's not entirely true. I can say that he is likely somewhere west of the Rockies. Next question."

"I'll rephrase that. What did you do to Walter?"

"Walter became a corporate problem, Lottie. You know at Veregro we tend to handle our problems a bit aggressively. I'm sure you remember. You used to carry the stick."

"Walter was a decent man. He would have helped you reshape Veregro into a company doing the right thing. The compassionate thing. He wasn't threatening you. He tried to enlighten you. But instead you sent your goons after him."

"Enlighten me? Walter went crazy. I pitied the man. He lost everything he cared about and he needed something to blame. It's too bad he chose Veregro. He needed to be shut down so I shut him down. And I can't tell you where he is. Not because I won't, because I don't know. Whatever happened to Walter is a tragedy, but he brought it on himself." Errol rested easily back in his chair and put his fingertips together in a triangle shape, signaling he was calm and in control.

"How can you sit back while the world is decaying because of Veregro?"

"Jesus Christ," Errol said. "Why am I surrounded by crazy people? Veregro feeds the world, Lottie. Did you ever think about how many lives I've saved? Poor Walter, woe is me." Errol crossed his hands on his chest, mocking Lottie. "Is that the extent of your world view? What happened to Walter? Veregro. Feeds. The. World," Errol pounded his fist on the table with each word.

"Veregro is killing the world, Errol. You know it. Alice knows it now, too. And Maya. And Sylvia."

"There's nothing to know," Errol shrugged.

"There'll be no fourth generation because of Veregro's treated GMOs. You caused MODS. Your reward to uncover the cause of IIS is a smoke screen. You're the reason your wife could never conceive. Walter and Randy's tests prove it all. And I know you know it. You went to Peru to find a surrogate. Really, Errol? Isn't that a bit of a coincidence? One of the few countries that have banned GMOs. It's just a matter of time before Veregro goes down. And you with it."

"Tell it to the FDA, Lottie," Errol said. "They've approved every seed we've shipped. Every patent. Every process."

"Because you've paid off everyone there," Lottie shouted.

The office door creaked open and Rafi poked his head inside.

"Sorry to interrupt. Can we just grab our tablets? We left them here earlier." Lottie nodded and Rafi and Todd came inside.

Errol's head whipped around to the door. "Todd? My boy! How's your mother?"

Todd froze and all eyes were on him. Lottie's mouth dropped open, then Simon's. Errol began to laugh and Todd ran out the door.

In the dark quiet of the birthing hut, Sylvia sat on the birthing stool set against the wall of the small room, and breathed. She could hear Felicia and Samarie in the next room, whispering quietly. But, after all this time, she was only two centimeters more dilated, and as much as she wanted to believe Felicia and not be afraid, she could feel fear's cold breath in the room with her.

Now she heard the outside door open, along with the momentary sound of the rain, which never seemed to end, before the door shut again.

"How is she?"

Alice! Sylvia had been waiting for Alice ever since Felicia and Samarie had brought her here. She was about to call out to her when Felicia spoke, answering Alice's question.

"Sylvia's not a big woman," she said, her voice so low Sylvia had to strain to hear. "And her labor's been going on a long time, but she's still not fully dilated. I was just talking to Samarie about what we might do to ease the baby along. I prefer everything to take its natural course, but in this case, if it goes on much longer, there could be some fetal distress."

"Is there something wrong with the baby?" Alice asked, and Sylvia could hear the alarm in her voice.

"Once labor begins, the only way to monitor the baby is to insert a fetal monitor. Hospitals do that. Doulas don't. But yes, I am concerned. The fetus isn't cooperating as it should be."

"Why do you keep saying 'fetus'?" Alice's voice was high and tense.

"It's what we call babies before they're born," Felicia answered in a calm voice.

"I want to see her!" Alice said, and a second later she was there, in the small dark room with Sylvia.

"Sylvia," Alice said, squatting so that she could place a hand on Sylvia's damp brow. "How are you?"

Sylvia tried to smile. "I am very tired," she said.

"Of course you are. You've been working very hard."

Felicia and Samarie followed Alice into the room. Samarie squatted next to Alice in front of the birthing stool, and Felicia went around the other side.

"Do you feel an urge to push?" she asked Sylvia.

Sylvia shook her head. "I already tell you no. I do not feel this."

"I'd like to examine you again," Felicia said. "Can you lean back and open your legs for me?"

Wearily, Sylvia nodded. "Maybe I will take the nap then," she said. Samarie smiled at her joke and patted her knee. Then, with Alice supporting one arm and Samarie the other, Sylvia spread her legs.

Each step of the internal exam, Felicia told her what she was going to do before she did it, so that none of her movements took Sylvia by surprise. Until, that is, a sudden pain stabbed through her and a small scream escaped from her lips.

"What is it?" Alice asked, her face a map of concern.

The pain was gone as quickly as it had come.

"The baby hasn't dropped at all," Felicia said, removing her hand and discarding her glove. She moved closer to Sylvia. "Sylvia, I'm going to do some manual external massage to try to move the baby into the birth canal. But I'd like you to sit straight up again. Are you ready?"

"I am ready for this baby to be born," Sylvia told her.

"You and me both," Felicia agreed as she and Samarie helped Sylvia sit more upright. Then Felicia pulled her own stool over from the corner of the room, sat down, and began precisely and smoothly massaging Sylvia's swollen belly.

Samarie had taken Sylvia's hand. Now she squeezed it. "Breathe into it," she whispered to Sylvia, as if they were the only two in the room.

"I am so tired of breathing," Sylvia said.

Alice, who held her other hand, brought both of their hands to her mouth and then kissed Sylvia's. "Breathe, goddamn it," she told her. "Breathe, or I'll breathe for the both of us. For all three of us. You got that? You breathe!"

Sylvia breathed, in rhythm to Felicia's long strokes. If she was still having contractions, she could no longer differentiate between them and the time between them. She closed her eyes. She felt something shift, deep inside of her.

"Good, Sylvia," she heard Felicia say, as if from far off. "Now can you push?"

Sylvia groaned, a horrid sound she could barely believe had issued from her own mouth. Then she pushed.

"Good! Good! Keep pushing! Good girl! Good! There...here...comes..."

Sylvia felt the baby slip from her. The room was suddenly silent.

"Samarie, begin compressions," Felicia said, her voice now sounding very near. Something else began to slide from Sylvia. She let it slide. She didn't care.

"Nothing," she heard Samarie whisper.

"Here. Let me try," Felicia said.

Sylvia opened her eyes and saw the lifeless baby as it was passed from Samarie to Felicia. *"Es muerto,"* she said, because she knew.

"No," Alice said, squeezing her hand. "Felicia and Samarie are giving him CPR. Everything will be fine."

"Es muerto," Sylvia repeated. Her son was dead. Only he was not her son. He was Mr. Errol and Ms. Alice's son. Poor Ms. Alice! "Ms. Alice," she said. "I am so sorry."

Alice didn't respond. The only sound in the room was the odd *whumpf* of flesh against flesh as Felicia tried to resuscitate the baby. Then that sound stopped as well.

"Es muerto," Sylvia said one last time.

"Would you like to hold him?" Felicia asked, holding the blue baby out to Samarie so she could wrap it in a blanket.

Sylvia turned her head away. "No," she said.

"Alice? Would you—?"

Tears coursed down Alice's face. "It's all Errol's fault," she said. "I never should have trusted him."

Felicia stepped toward her, held out the now-wrapped bundle. "Would you like to hold the baby?" she asked again.

Alice looked at the small blue face, practically buried in the bunting. *Es muerto,* she thought. But then something stronger kicked in, and she held out her arms, then held, for a moment, her stillborn son.

Twenty Seven

PETE WANDERED ACROSS THE PATH from Simon's office to pick a grapefruit. It was so big and heavy, the branch that carried it was bent down and the fruit nearly touched the road. Just as he snapped it off the tree, Pete heard the fast pace of footsteps coming from the office porch. He spun around and saw Todd leap off of the porch with Rafi bolting after him.

"Hey, what's going on?" Pete yelled as he instinctively ran after them. They were heading toward a field surrounded by a low wooden fence that kept the cattle from wandering onto the path. Todd used both hands to propel himself over the fence but Rafi leapt straight over it like a hurdling Olympic medalist. Rafi caught up, reached out and grabbed a handful of Todd's shirt, bringing them both down, rolling in the long grass and mud before coming to a stop with Rafi on top of Todd.

"You're working for Errol Foster? You son of a bitch. How could you do that?" Rafi screamed in

Todd's face. "This place took you in. Everyone here trusted you. What the fuck, Todd?"

"I'm sorry, Rafi. I'm your friend, man. Try to understand."

"Understand what? That you're a liar? That you betrayed Lottie and Samarie and the whole Orchard? That you used me to get in here?"

"Try to understand that my mom is dying. I needed to help her, man. Our house is in foreclosure, my dad couldn't make car payments. We didn't even have a car to get my mom to the doctor anymore," Todd screamed back. "We need medicine for her. I needed help. So yeah, I took Errol's money. What else could I do? What would you have done?"

Rafi got off of Todd, stood up and turned away, wiping his muddy hands on his pants.

Todd got up. "Rafi, can you understand?"

"Just go, Todd. Run."

Todd took off running and Pete, out of breath, came up to Rafi. "What the hell was that about, Rafi?" Pete asked.

"Todd was the mole," Rafi answered.

Lottie gave a little laugh. "Well, Errol. You got me again. I thought I could reason with you. I thought maybe I could appeal to your humanity somehow. But I'm swinging at air. I was a fool to think we could talk this out."

"Well maybe we can finally understand each other," Errol said. "You are correct: you're a fool. You're all fools. Not you, not Walter, not some fucking reports are going to stand in my way. I told you I will always be a step ahead of you, and here we are." Errol spread his arms wide. "You trust people too easily, Lottie. It clouds your judgment. So now you'll have to excuse me. I have a petition of eminent domain to file. Game over, kids."

"I think it's time to start the show, Simon," Lottie said.

Without a move from either Simon or Lottie, the lights went out, screens dropped to cover the windows, and the image of a smiling, naked man holding a spatula projected onto the long, white wall in the darkened room.

"What is this? Some sort of a sick joke?" Errol said.

"This is John Taylor," Simon said. "He's a customer of yours."

"Good, I like him already. Though I'd like him better in pants," Errol snorted a laugh at his own response.

"He runs a website called Peekaboo.com," Simon explained. "Foster Media recently acquired the internet provider that hosts his site. The previous owner said it was a hostile take over. But that's another story."

"Business is business," Errol said. "Make your point. I'm ready to see my wife."

"Taylor's a nudist. He has five webcams in his small apartment that are all working twenty-four hours a day. He's got four-thousand members paying fifteen bucks a month just to peek into his life. They can watch him cook in the nude, exercise nude, watch him shower, sleep, whatever. It's amazing what people will pay for, don't you think?"

"God bless America," Errol said.

The image on the wall changed. It was a bit grainy, but showed a parking lot adjacent to an apartment building.

"You recognize this building?" Simon asked.

"Should I?" Errol questioned back.

"Maybe," Lottie stepped into the dialogue. "It's in Omaha. It's the corner of Jones and 10th. It's the view outside the window of John Taylor's apartment, courtesy of his living room webcam. It's the building where Randy Hall lived."

The next few images on the wall came in quick succession as Simon spoke again. "As it turns out, Taylor keeps an archive of his video footage. So let's take a look at what we can see out the window the morning of March twenty-ninth, earlier this year. Here's a black Escalade that's driving by Randy's building. Thirty minutes later, there's a black Escalade in the parking lot. Let's check a close up, real quick, of this one. It's pretty grainy at this size but I have some software that can improve that. In just a sec I think we'll be able to read the license plate on the Escalade." The pixels of the image shifted and sharpened, until the image was clear. "Bingo, there it is. I love technology," Simon said.

Errol shifted uncomfortably in his chair and tapped his fingers lightly on the table.

"Now, I'm no detective, Mr. Foster. But there's a strange coincidence in play here. This license plate that you see, it belongs to a car registered to Craig Morgan." Up came a screen shot of the DMV record of the car's registration. "Veregro records show Craig Morgan has been an employee of yours for about twenty years. Hired not long after he was fired from the DEA for stealing drug money, and worked for you until, well, I guess until he disappeared earlier this year. Seems Craig paid a visit on the day Randy was killed. Let's quickly check the coroner's report, just to make sure we have our facts straight." Up came the coroner's report on the wall. "Yep. Same day."

"Morgan's been the head of my security for a long time. Randy Hall was suspected of compromising classified Veregro files. So Morgan was checking him out. Big deal. You don't have anything on me with this little slide show. Are we done now?"

"Not yet. That wasn't the first time Morgan did a sloppy job," Lottie interjected. "We saved the worst for last."

The next image that came up was a handwritten note with a signature on the bottom: John Angell. The suicide note.

Simon picked up a laser pointer and aimed it at the screen. "The signature itself is not a half bad forgery. But here, and here," Simon moved the laser to indicate different spots in the signature where the pen appeared to skip. "This shows where the author of the note stopped writing and then started again. Pretty uncommon for a signature, but not so uncommon when you trace something. Something like a signature. But that's not really the most compelling part. The writing above the signature," Simon was making a circle with the laser around the text above the signature. "According to the CIA's classified handwriting analysis software, all this here is a perfect match of a former government employee's handwriting. And lucky for us, Craig Morgan's handwriting hasn't changed much since he applied for a job with the Drug Enforcement Agency twenty-five years ago." Up on the screen came the handwritten portion of Craig's DEA application. "Amazing that the government digitizes all this stuff. Don't you think, Mr. Foster?" Simon smirked.

"God bless America," Lottie added.

Errol puckered his lips and gave a slow, polite clap of his hands. "Nice show. Really, I commend you. So I suppose you're going to ask me for something now. Right?"

"I want you to go public with Walter and Randy's report. Stop using Veresate," Lottie said. "Reformulate the Veregro seeds and if you can't, then shut down production. Help the farmers whose land you either raped or infected. Dedicate a few measly billion to making the world's farms healthy again. Dedicate funds to MODS research. Fight the disease. I wish we could do it on our own here, but the Orchard is just the model. We don't have the manpower and we don't have the funds, but you do. So that's our offer. Do it. Or we go public with this information you just saw, and overnight, Errol Foster goes from the man who feeds the world to a first-degree murderer. What's it gonna be, Errol?"

Errol drummed his fingers on the table again and stood up. "No deal, Lottie. But I'm proud of you. I didn't think you had it in you. When you want to do something big, you can't always play by the rules. You've got to use everything you've got. Everything. You did learn something from me after all."

"Tune into the six o'clock news tonight, Errol. Because your face is going to be all over it," Lottie said.

"Maybe so. And where will you tell them you found this information? You're holding stolen evidence from a crime scene. And you're the one who stole it, for Christ's sake." Errol gave out a deep laugh. "Sounds like quite a story to me. Angell Farm revisited. Makes a nice headline, doesn't it? Tell you what, let's race. So who can get who in the news first."

"You're bluffing, Errol."

"I don't need to bluff," Errol said. "As usual, I've got the winning hand."

Pete and Rafi were walking back to the office when they saw Errol step outside the door and head down the porch steps.

"I'm getting too old for this shit," Pete said as he broke into a trot, and Rafi followed behind him.

Lottie stepped outside the door and waved her hand to Pete. "It's okay, Pete. Let him go," she called out. Pete waved back as he and Rafi headed back to the office.

"I'm so sorry, Lottie," Rafi said as he went up the porch stairs. "I had no idea about Todd."

"It's okay, Rafi. It's not your fault," Lottie said.

"But I brought him here. I feel like it's all my fault," Rafi said, shaking his head.

"If it wasn't Todd, Errol would have gotten information some other way."

"Todd's been trying to take care of his mom," Rafi explained. "His family's broke trying to pay doctors and stuff."

"That's how Errol works," Lottie said. "He takes advantage of people's weaknesses. I should have been better prepared."

"But Todd, of all people. I just thought he was stronger than that," Rafi said.

Lottie took a step back from Rafi and put her hands on his shoulders. "Listen to me, Rafi. We can't understand what Todd's going through. In his mind, he's done the right thing. He did it for his family. Sometimes when people do bad things, there's a good reason behind it. Other people looking in, they only see the bad. But it's not always that simple. Life gets complicated. Sometimes people do bad things and keep secrets, but just because they're protecting what they love."

Rafi forced a smile. "Thanks for not hating me."

Lottie smiled back. "Go get yourself cleaned up, Rafi."

Rafi shook his head. "Maybe I better go check up on Samarie."

"You go, Rafi," Lottie said, giving him a pat on the back.

Lottie and Pete stood on the porch of the office, staring into the rain clouds. Simon came out of the office and looked at Lottie.

"So what's next?" Simon asked.

"I'm not sure, Simon. I'm really not sure yet. Maybe Alice will talk some sense into him."

"What about Errol?" Pete asked. "We can't just let him wander around the Orchard."

"No, we can't. Keep an eye on him. He said he's leaving after he sees Alice and the baby. Let's make sure that happens."

"Will do," Pete said, as he followed off in the direction after Errol.

"Do you think Errol is really going to the press with any of this? Or filing that petition?" Simon asked Lottie.

"I don't think Errol wants any bad publicity, even if he can spin it. If I know him like I think I do, he'll wait to see what we do first. If we strike, he'll strike back."

"With everything he's got," Simon added.

"Right," Lottie said. "The ball's in our court right now. I just don't know what the hell to do with it."

Errol went back to the guest cottage where Alice was staying but no one was there. He tried to reach her on her tablet but it went straight to her voice mail. She must be at the birthing hut, Errol figured. He was about to head out to find the place himself when the door opened.

It was Alice, wearing someone else's too-big rain slicker and mud boots; her hair hung stringy and uncombed beneath the slicker's hood. Errol leapt up to hug her despite her appearance, but Alice just stood there, arms at her sides. The slicker dripped onto the floor at her feet. "I was just about to go looking for you," Errol pushed the hood back and unsnapped the slicker, stepped behind her and tugged it off then hung it on one of the pegs by the door. He pointed to the bench beneath it and Alice sat obediently while he tugged off the mud boots. Then he pulled her to a stand and led her to one of the chairs by the woodstove. She sat again, then leaned forward to warm her hands.

Errol pulled the other chair closer, then reached for her hands.

Alice pulled them away.

"I can help warm them up," he said in a little boy voice.

"The fire can warm them up fine," Alice said, her voice already stronger.

"Where were you?" Errol asked.

Alice didn't look at him. "Walking."

"In this rain? Why?"

Now she turned, and studied his face. "You knew about the treated GMOs, didn't you, Errol? No. Don't lie to me. You knew. And you didn't tell me. Why? Why didn't you?"

Errol tried again to take one of her hands, but Alice slapped his away. Errol was growing angry, especially with the news he had just received, but he kept his composure.

Alice was watching him. "Errol," she said, her voice stern. "Why didn't you tell me?"

"I didn't want to lose you," he said in a small voice.

Alice laughed wildly. "Lose _me_? A lot more people's lives are at stake than just mine, Errol. How could you withhold such a thing? And why?"

Errol shrugged. "The tests were inconclusive."

Alice squinted at him. "Not the tests I heard about," she said.

Something hot suddenly burned at the center of him. His resolve returned, and he sat straighter. "Who told you about those tests? Lottie Winters?"

Alice nodded. "They're—"

Errol interrupted. "They're biased. They've been performed by only two lab techs. Veregro has years of testing, by hundreds of award-winning scientists, that countermand the tests she's citing. We have the FDA's approval. There is no definitive evidence that Veregro's products are causing MODS."

Alice had watched him throughout this speech. "You finished?" she asked.

"If you are," Errol said.

"Oh, I'm finished, all right. If you're going to sit there and lie to me like that, you can bet I'm finished."

This time, Errol's attempt to capture her hand succeeded. He held on. "Why would I lie to you, Allie?"

Alice tried to pull her hand away, but Errol held it tight. "I don't know, Errol. Why would you lie to me? That's what I've been walking around asking myself. Guilt? Maybe it's too much for even you to handle. That you're the reason I could never carry a child."

Errol didn't respond.

"I've always defended you, Errol. I know all about what other people say about you behind your back. And I know the way you rule your companies, the way you won't let even someone like Maya have too much control. But I always thought you and I were above all that, Errol. I thought I knew the real you. I thought that, because you'd shown me the frightened little boy in you, what you and I had was different from the Errol Foster the rest of the world saw."

"It is," Errol insisted.

"Look at me," Alice said.

Errol looked at her.

"Don't lie to me," she told him.

"I won't," Errol said.

"Do treated GMOs cause MODS?" she asked.

Errol looked away. "No," he said, blinking wildly. Now was probably not the time to mention that he himself hadn't eaten anything that might contain treated GMOs in years.

"I don't believe you, Errol." Alice shook herself free of his hand and went to the door. She slipped the mud boots back on and slid back into the still-dripping slicker.

Errol tried to stand, but seemed rooted to his chair. "Don't go," he said. "We should be celebrating now. We can talk about these accusations at a better time."

"Celebrating?" Alice huffed. "Celebrating what?"

"Our baby, Alice! I'm here because the baby is coming. We're going to be a real family now, so let's enjoy the moment. I'm getting us out of this shithole and finding a comfortable place to stay. We'll get a nurse and a nanny, whatever we need."

A tear streamed down Alice's face. "You didn't get your heir, Errol. The poor baby didn't make it, and you only have yourself to blame. I want you to leave and I'll be staying here."

Errol balled his hands into fists and stood up. Control was slipping away. "What happened to the baby?" he asked through gritted teeth.

"He was stillborn. Sylvia hasn't felt a kick for two weeks. And she was afraid to tell us because she's terrified of you, Errol. She's smarter than me, that's for sure."

"Get your things together, Alice. We're leaving this place now."

"I told you, Errol. It's over. I'm not going anywhere with you. I'm going back to Sylvia; she needs me right now." Alice opened the door but Errol darted toward it with his full weight and slammed it shut.

Errol gripped Alice by the shoulders. "Forget about that lying bitch, Alice. You're coming with me."

"Get your hands off me," Alice yelled. "You're crazy." She struggled to pull free from his grip but he was too strong, so she scratched at his face with her fingernails and Errol flung his head from side to side to avoid her attack. Errol let go of her with one hand to protect his face, but he knew he had to end this madness. Still jerking his head back and forth, Errol let go of Alice and grabbed the lamp sitting on a table in the entry. Alice reached for the door but one strike to the back of her head dropped her to the ground.

TWENTY EIGHT

ERROL SLIPPED OUTSIDE the door of the bungalow and walked to the SUV parked out front. He opened the driver side door and checked under the floor mat. The PA who picked him up from the plane left the spare set of keys precisely where Errol had instructed. It wasn't the first time Errol anticipated that he might need to leave somewhere in a hurry, so he was used to being prepared. Lottie's goon Pete was watching him from across the road, arms folded across his broad chest.

"Are you going to open the gate or shall I just drive through it?" Errol called out to him.

Pete said nothing, just shook his head and started down the road toward the gate. Perfect, Errol thought. He just needed Pete distracted for a minute so he could load his precious cargo. Errol rushed back into the bungalow and lifted Alice up. She was out cold and wrapped up in the blanket from the bed. Errol carried her outside, carefully laid her down in the back seat of the SUV and covered her face with the blanket. He hopped into the driver's seat and headed for the front gate. All three of Lottie's goons stood at the gate and watched Errol drive through.

"Glad to see that guy leave peacefully. I thought he was gonna be trouble," Pete said to Shooey and Rev.

"Well, amen to that," Rev said as he walked the gate closed.

In the ten minutes it took Errol to reach the plane, the rain started to fall harder than before. He wasn't concerned, though, the flight to Santa Barbara was just a short nautical hop. They'd be in the air less than ninety minutes.

Errol had to throw Alice over his shoulder to lug her aboard the jet. He placed her down on the couch so he could unfold one of the chairs that converted into a small bed, then he transferred her over and strapped her in. In the cabinet behind the cockpit, he hung up his coat and grabbed a towel to dry off his hair. Just one more task before settling into the cockpit. He reached inside a console that sat between two of the lounge chairs and pulled out a tumbler and a bottle of scotch. He poured himself some, swallowed it in a single swig, then refilled the glass. Now he felt ready.

Once in the cockpit, he switched on the power and checked his gauges. His fuel level was still good. He ran through his checklist, ending with the flaps on each side. He strapped himself in, released the brake and checked the engines, one at a time. He flipped on the radio, prepared to check in with the tower at Oakland Airport.

Errol taxied to the far end of the field. The wipers were useless but he left them on anyway. Then he pushed the throttle forward and began to accelerate, gaining speed. He pulled back the throttle and felt a rush, a warm tingly feeling he always felt when his jet left the ground. Once he reached cruising altitude, he entered navigation settings and flipped on the autopilot. He took another sip of scotch and felt like he was already home.

Lottie and Samarie sat on the porch of the birthing hut. "Errol never even wanted to meet me, did he? I don't understand why he was so anxious to know if I was really his if he never even wanted to see me," Samarie said.

"I'm sorry, Samarie. I didn't make it a very comfortable visit for him. There was some business we had to discuss and it didn't go as smoothly as I had hoped," Lottie explained.

Pete came jogging up and ducked under the overhang to get out of the rain.

"Been looking for ya," Pete said. "Our guest has left. I watched him drive out the gate myself."

"Thanks for keeping an eye on him, Pete. We heard his jet take off a few minutes ago. Did you happen to see Alice? She left here a little while ago and we figure she could use a friend right about now."

"I saw her and Errol back at her cottage before he left. She must have told him off pretty good because he left alone and didn't look too happy."

"She asked Samarie if she could stay here a few more days. She's a strong woman. She's been through a lot today." Lottie gave Samarie a pat on the knee.

"Let's go pay her a visit. See if there's anything she needs." Samarie nodded. They zipped up their coats and headed for the guest cottages.

Lottie knocked on the door of Alice's cottage but no one answered. She waited a minute then knocked again, but still no answer.

"You don't think she would do anything stupid, do you, Mom?" Samarie asked.

"You mean like hurt herself? I seriously doubt it, Sam."

Samarie walked over to the window and peered inside.

"Something's not right, Mom. There's a lamp on the floor and the bed is all messed up. It looks like there is a stain on the bed or something. It's hard to tell."

Lottie came over to the window and looked inside. Lottie knocked again, hard enough to shake the door, but she didn't wait long for a response this time. "Wait here, Samarie. I'm going to get the spare keys."

Lottie came back with Pete and they unlocked the door. Lottie went inside first and called out. "Alice? Are you here? Sorry to come barging in, but we're worried about you."

No answer. Pete came inside and picked up the lamp. The prongs on the plug were bent to the right. Lottie walked over to the bed and looked at the stain. She touched it; it was still wet. "This looks like blood," she said to Pete.

Pete was still holding the lamp and came to look at the bed. "Something bad happened here, Lottie."

Errol turned on his tablet and sorted through his new emails. He clicked one from Marcia titled: DNA Results. "Fuck," he said to himself after reading it. He swiped away the emails, activated the phone, and said "Marcia" to the tablet. After a moment she answered.

"Hello, Mr. Foster. What can I do for you?"

"Get Rick Fisher on the line. I'll wait."

"I'll try to reach him, sir. But he's on vacation this week."

"Reach him, it's an emergency," Errol instructed.

"One moment, sir." Marcia was gone for a moment and returned. "I am unable to reach him, Mr. Foster. It went straight to his voice mail."

"That is unacceptable," Errol screamed. "Who authorized this vacation?"

"I believe you did, Mr. Foster."

"Well the authorization is revoked. Find him and tell him to get his ass back. Now. We need to make changes to my will and they need to happen immediately. Is that clear?"

"Crystal, sir."

Errol disconnected and dropped the tablet on the empty seat next to him before slamming the rest of his scotch. A moment later he heard the cockpit door open.

"Alice, honey. I'm sorry I hit you but I had to get you out of there so we could take a moment to calm down and talk like adults. I hope you're feeling okay." Errol turned around in his seat to offer a smile, but it wasn't Alice who was standing in the doorway. It was Walter Conroy.

Lottie ran to Simon's office and threw open the door. "Simon, I need your help."

"What's happening, Lottie?" Simon was shaken by her panicked state.

"Errol just took off in his plane. I think he took Alice against her will. She needs help." Lottie said. "Can you find out where he's headed?"

Simon tapped a few keys on his keyboard and up on the screen in front of him appeared a map of the western coast of the United States next to a mass of blue; the Pacific Ocean. Simon began typing some more and the map got larger. Soon it was just California in the view. A moment later a small green blip appeared, blinking on and off and heading south. Simon pointed to it. "I think this is him," Simon said.

"Are you sure?" Lottie asked.

"I can't be a hundred percent certain, but it's the closest plane to us right now and it's not on an official flight path. That's a sign. Also, it's moving faster than a commercial plane but not as fast as a military jet, so it's likely a small, private jet. I think it's him, Lottie."

Lottie nodded. "He's probably headed home, to Santa Barbara. Let's keep an eye on that plane, Simon. Once we know where he's going, I'm calling the police."

Walter held a Glock 22 aimed at Errol's head. "Surprised to see me, aren't you, Errol?"

"What are you gonna do, Conroy? Shoot me? Then we both die."

"Maybe, but you die first," Walter said.

"You're not going to shoot me, Walter. If you had the balls, you would have shot me already on the ground."

"I'm not stupid, Errol. You have too many resources on the ground. Your hit squad is probably hiding in the bushes. Up here, I know we're alone. No one's coming to help you. So we can talk, and maybe we'll make some progress."

"And if not?"

"Then I'll have to shoot you."

"Okay, Walter, let's talk. Why don't you relax? Grab some scotch, take a seat up here with me and we'll talk, like civilized men."

"I'll stay right here. Whether or not we stay civilized, that's up to you," Walter said.

Errol let out a deep laugh. "What a tough guy you've become, Walter. You outta be thanking me. You're a better man now."

"Cut the shit, Errol. There's an opportunity for you here. No one has to die. You can do the right thing and take all the credit. Think about it. Errol Foster feeds the world, now Errol Foster saves the world. That's what I was trying to get through your skull when I first came to see you with the reports. Before you tried to have me killed."

"When you first came to see me, you were crazed, Walter, like a rabid dog. There's only one way to deal with a rabid dog. You can hardly blame me."

"I was in mourning but I wasn't crazy. The reports are real, Errol. They're solid. Veregro caused MODS."

"The report is almost twenty years old, Walter. You think the world is just going to forgive me now, invite me to help solve a problem I've known about for twenty years? You're still fucking crazy," Errol said.

"No one knows how old the reports are, Errol. That's between me and you. And Randy. But I'm willing to forgive his murder for the greater good. You can go public with it today, like it was just discovered. No one needs to know. Just do the right thing."

"Oh, how gracious, Walter. You'll forgive me so I can endure the lawsuits and the public witch hunt. I'll be stripped of billions before I can help a soul. I won't be able to feed my own family, even with the goddamned poisonous food we grow. I'll starve and everyone else will still die."

"People are desperate, Errol. You'll be a hero."

"I wish I could make you understand my view of the world, Walter, but this flight just isn't long enough."

Errol hit the center yoke and the airplane twisted onto its side and banked sharply to the right, sending Walter off balance and to the floor. Errol hit the yoke again to the other side. The plane twisted to the left and Walter rolled across the plane, dropping his gun. He tried to get up and find it, but there was no way to keep his balance as the plane rolled from side to side.

Errol looked back and laughed when he saw Walter on his back, rolling around. He steadied the plane, re-engaged the autopilot and jumped up to close the cockpit door. Walter was able to slide his foot just inside the entry and stop the door from shutting. Errol slammed the door against Walter's ankle over and over, and Walter howled in pain. Then Errol began to kick at the limp foot when he suddenly stopped, reached for something on the floor and came up with the gun.

Walter was lying on his back, in a wet puddle of scotch. He reached behind his head and felt something smooth and angular. Whatever it was, he grabbed it and hurled it forward with all his strength. The empty crystal decanter sailed right into Errol's face, who fell forward onto Walter, motionless.

"What was that?" Lottie asked.

"It looks like the plane made several sharp turns from side to side," Simon said. "It might have been a software glitch, it's hard to say. It looks like they're still headed south."

"Is there any way to hear what's happening inside that plane, Simon?"

"ATC frequencies are public so if they're communicating we should be able to pick them up." Simon opened up a window on the computer with a graphic of a ham radio interface and started scanning through channels.

"American five eight seven heavy wind three zero zero at niner runway three one left cleared for takeoff."

"Delta five eight seven heavy Oakland tower caution wake turbulence runway three one left taxi into position and hold."

"United sixteen eighty three wind three one zero at six runway three one left kilo kilo intersection cleared for takeoff."

"These all sound like commercial airliners, Lottie. We'll just have to keep listening."

Walter rolled Errol off of himself and sat up. He unbuttoned his shirt and looked down at the cell phone taped to his chest. Walter had recorded their entire conversation: Errol admitting knowledge of Randy's killing, the attempts on Walter's life and best of all, the truth of the Veregro reports. But none of that mattered now because his plan had backfired. He was just going to use the gun to make sure he could get away unharmed once they were on the ground. But he got cocky, tried to be threatening, and now he was the only conscious person on a plane forty thousand feet in the air.

He limped up to the cockpit and dropped himself in the pilot's seat. The plane was flying steady but rain was pounding the windshield. He looked around for a radio and located a control panel labeled COMM. It was turned on but he heard nothing. Then he spotted the head set resting close to the throttle yoke. Careful not to disturb the yoke, he grabbed the headset and put it on. All he heard was static.

"Mayday, mayday. Hello? Can anybody hear me?"

"This is Bay TRACON. We can hear you. Go ahead," said a female voice.

"I'm in a private plane. The pilot is..." Walter looked back at Errol lying on the floor. "Incapacitated."

"Do you know your heading and coordinates?" the controller asked.

"I'm not sure what you mean," Walter said, beginning to panic.

"Do you know where you are, sir?"

"Flying along the California coast, headed south. We left Marin County maybe thirty minutes ago."

"Okay, I think we've found you. What's your name, sir?"

"My name is Walter. Walter Conroy."

"Okay, Walter. Just try to stay calm. I'm Cindy. I'm going to do my best to help you. Do you know what happened to the pilot?"

"Um, he got drunk and passed out."

"Do you have any flying experience, Walter?"

"Zero," Walter said. No response. "Cindy, are you still there?"

"I'm still here, Walter. I'm just confirming a few things."

"I'm going to die, aren't I?"

"We need to work together here, Walter. Try to stay calm. I'm going to give you some instructions so please listen carefully. If you have any questions, ask me before you touch anything, okay?"

"Okay," Walter said, hands shaking.

"You're headed into some very severe weather conditions, Walter. We're going to need to turn you around."

Simon was still rolling through stations when he came across an odd conversation. He turned the volume up. "Lottie, listen to this. This isn't a commercial pilot talking," Simon said. They listened to the static filled conversation.

"I'm not sure what you mean."

"Do you know where you are, sir?"

"Flying along the California coast, headed south. We left Marin County area, maybe thirty minutes ago."

"Okay, I think we've found you. What's your name, sir?"

"My name is Walter. Walter Conroy."

Lottie and Simon looked at each other with blank stares. "Am I crazy or did that guy just say his name is Walter Conroy?" Lottie asked.

"He did," Simon said, looking confused.

"This doesn't make any sense," Lottie said shaking her head. They continued listening in fascination. Finally, Lottie jumped up. "I've heard enough, let's go."

"Where are we going?" Simon asked.

"Where do you think? Marin County Airport."

"Okay, Walter. We're going to reprogram the autopilot settings to head you to Marin County Airport. You're looking for LED displays with six sets of numbers. They should all be lined up in a horizontal row. The numbers will probably be blue."

"I think I found them," Walter said.

"Read me the numbers that you see, from left to right," Cindy said.

"34, 25, 34, 119, 50, 25."

"Were you heading to Santa Barbara?"

"I think so," Walter said, perking up.

"You're doing great, Walter. Those are your autopilot settings. There should be a dial under each of the displays. Now we're going to change the numbers and reset the system."

After reprogramming the autopilot, Walter felt the plane gracefully turn in a large arc until the ocean was on his left.

"We turned. It worked," Walter exclaimed.

"Now just relax. The plane will take care of itself for the next twenty minutes and after that we will be getting ready to land. Let me start preparing you."

Walter was repeating the instructions over and over in his head. Make sure the plane is level, correct the pitch, reduce power, pull back the throttle, but not more than a quarter of an inch. Oh, and look for the runway. Walter heard a loud whirring sound as he slowed the plane and then turbulence hit. The plane felt like it was bouncing up and down and Walter was fighting to hold the throttle steady. He looked out the windshield looking for the runway. He was going to have to make a sharp right turn to line up properly. He moved the yoke to the right and the plane started to turn onto its side. So he moved the throttle back but it was too much, too quick. The plane was now rolling toward the left and bouncing up and down. Panic filled him. The plane was out of his control and he wondered if he would die quickly when he crashed. He prayed that he would.

"My God, my God, my God," he was chanting when suddenly he felt a soft hand on his shoulder. An angel, so soon? He turned to look.

"Let me sit here," Alice said. "I can land this bird."

The Marin County Airport was used mostly for flight training and private planes. Cindy was smart to direct them there since there would be no other airplane traffic. When Lottie and Simon arrived, the emergency response crew was already in place, waiting. Through the clouds and the rain they watched the plane turning toward the runway and starting to descend. The plane was unsteady, pitching from side to side as it came down.

"They're going so fast," Simon said. "shouldn't they be slowing down?" Lottie didn't respond, she just stood and watched with her hand on her chest.

The plane hit the landing strip and the front end bounced up and down. On the third bounce the left end of the plane tilted up, crumpling the right wing against the tarmac before it ripped free of the fuselage. The plane skidded off the runway and barreled into a lot of small, single engine planes before it came to a stop with fire blazing from the wound where the right wing belonged.

The fire trucks and ambulances raced to the plane. As they got there, the plane door unfolded and two figures climbed down the steps, arms around each other. Two firemen reached them and nearly lifted them off the ground to quickly move them away from the burning wreck.

Simon and Lottie ran toward the ambulances, where Alice stood shaking her finger toward the plane.

The firemen were headed back to the wreck when it exploded, shooting flames higher into the cloud-covered sky, and the firemen backed away to retrieve their hoses.

Lottie grabbed Walter and hugged him tight. "I can't believe you're alive," Lottie said, starting to cry.

"Believe me, I'm as surprised as you," Walter said with a warm smile. "Alice got us on the ground. She saved us."

"How on earth did you end up on that plane?" Lottie asked.

"How on earth did you know that plane was coming here?" Walter asked.

"Touché," Lottie smiled.

"I'll tell you the whole story, Lottie. But I think we should get Alice out of here first."

"Alice, are you okay?" Lottie asked.

"I'm going to be," Alice said.

"What about Errol?" Lottie asked.

Alice looked toward the wrecked plane, and shook her head. "I told the firemen there was someone else in the plane but...."

"I'm sorry, Alice," Lottie said. "Despite everything, he was your husband. I wouldn't wish this on anyone."

"Yeah, my husband," Alice said. "He wasn't all bad. He taught me to be a pretty good co-pilot. Goodbye, Errol." Alice said, watching the burning wreckage.

Lottie looked back at Walter, pondered him for a moment and then kissed him on the lips.

"What was that for?" Walter asked.

"Because you're here," Lottie said with a shrug. And Walter kissed her back.

Twenty Nine

AFTER A SHORT VISIT to the hospital so Alice and Walter could be checked out, they were on their way back to the Orchard. Both Alice and Walter had minor wounds. Their crash landing resulted in less damage than the lamp to Alice's head and the door slams to Walter's ankle.

Lottie called ahead to say they were on their way home, and a group gathered in the dining hall. A cheer erupted when they arrived.

Maya, Ginny, Elena and Felicia surrounded Alice and pulled her close to the fireplace.

"We were so worried about you, Alice. Thank God you're all right," Maya said.

"What can I get you?" Ginny asked. "Do you want some tea, or cocoa maybe?"

"How about some pinot?" Alice said, and the ladies all laughed. "I'm not joking! Give me wine, dammit." Alice started to laugh along with them, and then burst into tears.

"Do you want to go rest, honey? You've been through a hell of a lot today," Maya said.

Alice looked around the room. "No, I don't want to be alone right now. This is what I need. The company of good friends. Is Sylvia okay?"

"She's doing fine," Felicia said. "She's sad though, and worried about you. She'll be happy to see you in the morning."

Rafi and Samarie both embraced Walter. "We thought we'd never see you again," Samarie said.

"I really missed you guys. I missed this whole place," Walter said, looking around. "I'm glad to be here."

"Will you come with me for a minute?" Lottie asked Samarie. They went and sat down at a table away from the rest of the group and the noise, then Lottie told Samarie about Errol.

"I guess I should feel sad because he's my dad, but I never even knew him," Samarie said. "I felt worse when Walter left. Does that make me a bad person?"

"No, it doesn't, sweetie. It makes perfect sense how you feel. I just wanted to be the one to tell you. I didn't want you to hear it from someone else, and it's going to be all over the news soon."

"Do you think anything will change with Veregro, now that he's, you know, gone?"

"I don't know, Sam. But let's hope something good will come of it." Samarie smiled slightly and nodded her head.

"You and that baby might want to think about some sleep now," Lottie said.

Alice found Lottie, hugged her and kissed her cheek. "I want to thank you, just, for everything. I owe you so much. You had my back. I'll never forget that. Sylvia and I will get out of your hair tomorrow."

"You don't need to leave, Alice. Neither does Sylvia. Stay as long as you want. You can even call this home if you want to," Lottie said. "I know a perfect spot for your garden."

It was getting late and the crowd in the dining hall thinned out as folks headed off to bed. Lottie sat at a table with Walter and Simon when she noticed they were the only ones left in the hall.

"Okay, who's ready for some of the hard stuff?" Lottie asked.

"Bring it on," Simon said.

"I'm game," Walter said.

Lottie went to a cabinet and came back with three shot glasses and a clear, unmarked bottle. "This is Pete's organic potato vodka. His ancestors are from Russia, and this recipe is what they brought to America."

Simon and Walter laughed while Lottie uncorked the bottle and poured three shots. They all raised their glasses and Lottie thought for a moment before making her toast. "To the men in my life. That's you, Walter, and you, Simon. My sweet J.P. And Pete, wherever the heck he is. And to my dad. If not for Charlie Winters, we wouldn't be sitting here right now," Lottie looked up. "I'm thinking of you, Dad. *Nazdarovya!*" Their glasses clinked and they all drank.

"It's time to spill your guts, Walter Conroy. We want to know how you ended up on that plane," Lottie said.

"It was a fluke really. What happened was I just got tired of running. Errol's guys were after me. No matter how often I moved they found me. I couldn't sleep anymore, I was always waiting for the sound of gunfire or my door to be kicked in. Everyday was consumed by planning where I would go next and then getting there."

"They were able to find you from the IP addresses used to make your blog posts," Simon added.

"Yeah, it took me a while to figure that out. I was using secure VPNs, so I assumed my tracks were covered."

"But Errol owns most of the VPNs available in the U.S., so he could trace you anyway," Simon said.

"I finally figured out what was going on when I couldn't access my site anymore. I was afraid to use my email or text anymore, afraid to even open them. I worried they could trace it somehow and find me," Walter explained.

"After your blog went offline and we never heard from you, we thought they found you. We thought you were dead," Simon said.

"Not dead, just paranoid. So I went to a pawn shop and bought a gun. I know that sounds crazy but I was prey. I was just waiting for the hunters to appear."

"It's not crazy, Walter. It's really scary, is what it is," Lottie said, reaching out to grip his hand.

"You're damn right I was scared. I didn't feel safe anywhere. I figured, I need to go off the grid completely until they stop looking for me. So I came here."

"You've been here?" Lottie asked.

"I've been camping in the forest for two weeks, outside the Orchard. Why not? There's lots of people camping around here, waiting for their chance to visit the Orchard, different people come all the time. I blended right in. I finally felt safe. Then today I saw Errol's plane land, and that's when my idea was born."

"To hijack his plane?" Lottie asked.

"Look, I know it was a half-baked idea, but my intention was to talk to him in a place where no one else could reach us. That's the only way I felt safe."

"That's the mistake we all made, Walter. Thinking we could reason with Errol Foster," Lottie said.

"Yeah, well I didn't have total faith in that approach. So here was my back-up plan." Walter set his phone on the table and played back the conversation he recorded on the plane.

"Errol knew all along that Veregro caused MODS," Lottie said.

"So now what?" Simon asked.

Walter reached for the bottle and poured another round of shots. "We wait to see who's in charge of Veregro now."

By morning, the rain had eased to a drizzle, though the sodden grasses misted so thickly it was difficult to see more than ten feet beyond the rail. Samarie and Rafi sat on the porch, legs propped on the rail, each with a cup of tea.

"Good morning, you two," Lottie said, poking her head through the screen door.

"Morning," Samarie and Rafi said in unison.

"I'm starting breakfast so if there are any special requests, shout 'em out now," Lottie said.

"Eggs," Rafi said.

"Bacon," Samarie added.

"Potatoes," Rafi said.

"With onions," Samarie added.

"Okay, is that it?" Lottie asked.

"Oh, and a bagel with cream cheese and lox and capers and pickles with chocolate ice cream and a cherry on top," Samarie requested, and they all cracked up.

"Yeah, that one's coming right up," Lottie said, still laughing.

"Uh-oh," Samarie said.

Rafi looked down. "What was that?"

"I think I just laughed so hard that I peed," Samarie said.

Lottie came running outside and looked at the puddle. "You didn't pee, Samarie. Your water just broke. The baby's coming!" she shouted.

Rafi leapt from the chair so quickly it bounced back. "It's time?"

"It's time, all right. You guys head to the birthing hut and I'll call Felicia," Lottie said.

"But, Mom," Samarie said.

"I know, I know. You're hungry. I'll bring you food."

Samarie and Rafi headed off into the mist in the direction of the birthing hut.

With Amanda holding one arm and Rafi the other, Samarie squatted, facing Felicia, who was checking her dilation. Yes, her contractions hurt, but, unlike Sylvia's, they didn't wrack her body or make her fearful. Her baby was coming. And all would be well.

Felicia withdrew her hand and smiled. "You're fully dilated," she said. "Push whenever you're ready, honey. This baby's ready to be born."

"So fast," Sylvia marveled. "This is good, Samarie. You and your baby will be fine. I know this."

Samarie squeezed Sylvia's hand. "I'm so glad you're here," she said. "I hope this will help you begin to heal."

"You help me to heal," Sylvia assured her. "Our hearts are woven now, sisters, together."

Suddenly, Samarie felt an overwhelming urge to bear down. She groaned, and then everything she'd been feeling changed. The pain was replaced by a glorious sensation of movement, of release, of—*oh my God!* She heard a cry. A baby. Her baby. Their baby.

"It's a girl," Rafi said, as tears streamed down Samarie's cheeks.

Felicia passed the baby to Samarie. The little face nuzzled into Samarie's neck. The little fists curled against her collarbone.

"Oh my God," she said, aloud this time.

Rafi stroked both Samarie's hair and the baby's head. "Our Hope," he said.

Felicia smiled. "I wondered if you'd decided on a name," she said. "Hope is lovely. Perfect."

"Hope Sylvia," Samarie amended.

"Oh, no," Sylvia protested. "You do not need to name your daughter with my name."

"Of course we don't need to," Samarie told her. "We want to."

The door in the other room opened, and then they heard Lottie's voice. "Here, Walter, you sit down. I'll go see how it's going."

"Hello!" she called before she entered the birthing room. Then she stopped, her eyes brimming at the sight of her daughter and her newly born granddaughter.

"Hope Sylvia Winters-Summers," Rafi said, "meet your grandma."

Careful not to disturb the bonding between Samarie and Hope, Lottie lay a hand on her daughter's shoulder. "You done good, kiddo," she said, her voice gruff.

Samarie smiled at her. "Thanks, Mom," she told her. "For everything."

Thirty

RICK FISHER WAS THE LAST PERSON Lottie ever expected to show up at her door, but there the Veregro corporate counsel was, hat in hand and looking contrite as a slime possibly could.

"Have you come to apologize?" she asked him, opening the door.

"Is Samarie here?" Fisher asked.

"What do you want with my daughter?" she snapped. "She had a baby three days ago and I don't think your company is what she needs right now."

Fisher held up a hand. "I'm not here to fight with you, Ms. Winters. In fact, I think you'll like what I have to share with Samarie. Can you get her for me? Please?"

Lottie offered him a squint of doubt, but settled him in the kitchen before going to Samarie's room and knocking on the door.

"Come in," Samarie called.

When Lottie opened the door, she found Samarie sitting up on the bed, nursing Hope. "Hey, Mom," she said.

"There's someone here to see you," Lottie told her. "Rich Fisher. Veregro's corporate counsel."

"To see me?" Samarie said. "That makes no sense."

Lottie shook her head. "No. It doesn't. But he insists on talking to only you, and the sooner I get the man out of my kitchen, the happier I'll be."

Hope had fallen asleep, and Samarie gently disengaged her, then stood to lay her in her cradle. Now she brushed at the front of her dress. "Do I look presentable?" she asked Lottie.

Lottie went and wrapped her in a hug. "You look gorgeous," she said. "You always look gorgeous."

"Okay, then," Samarie said. "Let's get this over with."

Walter had limped into the kitchen for a cup of tea, and nearly fell over when he saw Rick Fisher sitting there.

"What are you doing here?" he demanded.

"I'm here to see Samarie," Fisher answered easily.

Walter turned and got down a teacup, then he turned back to Fisher.

"Samarie?"

"That's what I said." Fisher busied himself with his tablet.

"Why?"

Fisher looked up again. "You of all people, Conroy, should understand attorney-client privilege."

"Samarie's your *client*?"

Before Fisher could respond, Samarie and Lottie came into the kitchen. Lottie went to Walter and wrapped him in a hug while Samarie shook Fisher's hand and sat down.

Fisher turned to Walter and Lottie. "Alone, please."

"Why can't they stay?" Samarie protested.

"It's confidential information," Fisher told her.

Samarie shook her head. "I don't have any secrets from my mother and Walter."

Fisher sighed elaborately. "All right, then. Lottie. Walter. Please. Sit down."

"I'd like a cup of tea first," Walter said.

"I'll get it," Lottie told him.

"Sam?" She paused for a beat. "Rick?"

Fisher looked at his watch. "Can we get started, please?"

Lottie put the kettle on. "Go ahead. It's a small kitchen. I can hear you."

Fisher frowned, but reached down and pulled a fat folder from his briefcase. He extracted two copies of a document, set one in front of Samarie and one in front of himself.

Walter didn't have to move to see what it was.

LAST WILL AND TESTAMENT OF ERROL THOMAS FOSTER.

"Lottie," he said. "You'd better come sit down."

Errol left all his business interests to his sole heir, as the will was worded. He had never intended that to mean Samarie; he had never intended to have more than a single heir, his newly born baby, but fate intervened. Foster Media. His majority holdings in Veregro and Deep Sea Exploration. His shares of Brazil Energy and the New Mexico Spaceport. Billions of dollars. And control over a great many things that mattered far more than all that money.

Samarie pushed her chair away from the table and stood up.

"You need to sign this," Fisher said, holding out a pen to Samarie.

She took the pen, signed the document, and walked away in a daze, saying nothing.

It had been a week, and Samarie hadn't said a word about her inheritance or the immense responsibility looming over her. She was solely focused on Hope and Rafi. Her family. Lottie was so proud of Samarie, she couldn't see her without breaking out in a foolish grin.

Finally, she asked Samarie to walk up the hill with her. Around them, the wildflowers had burst into bloom. Carrying Hope easily in her sling, Samarie walked next to Lottie until they reached Lottie's aerie and settled against two sun-warmed rocks. "You probably know I've called you up to my office for a reason," Lottie told her.

Samarie laughed. "Will my office be as nice as this?"

"The world is truly your oyster now," Lottie said. "Thanks to...your father."

Samarie sobered. "I never got it, Mom. You and Errol. Why?"

Lottie watched the baby nursing for a moment. "It's all wrapped up in a story I never told you. About what happened when I worked at Veregro."

Samarie nodded. "I know about that. Angell Farm. It wasn't your fault."

"I know you've read what's online and what's been on the news, Samarie. But there's more to the story."

"It's about J.P. I adopted him after I left Veregro," Lottie said.

Samarie couldn't disguise her surprise. "J.P's *adopted?*"

Lottie nodded.

"Well, what does that have to do with Angell Farm?"

"Because J.P is John Paul Angell. He's the boy that survived that massacre."

Samarie had been watching Hope, but now she looked up. "Holy fuck. Excuse me for my language. But holy, holy fuck, Mom. Poor J.P. Does he know?"

Lottie shook her head. "He has no memory of what happened. I took him to a psychiatrist, down in San Francisco, years ago. I was worried that he might have PTSD, as so many of the vets here do, or that he might recover the memory years later, and then blame me for not telling him. But apparently the memory was stored in the part of his brain that no longer functions. It's a strange sort of gift, but it's a gift nonetheless."

Samarie returned to watching her daughter nurse. "J.P not only adopted, but the surviving Angell. Jeez, Mom. You got any other bombs you want to drop?"

"I think that covers it," Lottie said, gently stroking the top of Hope's head.

"Can I ask you something?" Samarie asked.

"Yeah, I think that's fair," Lottie smiled.

"In all the news stories about you, they call you Charlie Winters. How come you decided to change your name?"

Lottie sighed. "That's a good question, Samarie. My parents always called me Charlie, short for

Charlotte. That was my name growing up. And I loved it. My dad was big Charlie and I was little Charlie. It broke my heart to stop using that name. But after I was scapegoated for what happened to the Angells, I didn't want anything to do with Charlie Winters, ever again. I managed to adopt J.P. I needed a fresh start, to distance myself from the stories. That was all that mattered."

Samarie shifted Hope to her other breast. "Well, for whatever it's worth, I think Charlie Winters is a beautiful name."

Lottie plucked a yellow wildflower, spun it between her fingers and looked at Samarie and Hope. "Yeah, so do I."

Samarie had never flown before. Rafi, Hope, Lottie and Walter accompanied her on her first trip to Omaha. Looking out the window of the plane, she realized she loved seeing the Earth from this angle. It made you realize just how fragile it was, how small.

Once at Veregro headquarters, Samarie prepared for a meeting with the board and the major shareholders. "You're sure about this?" Lottie asked.

"I'm sure, mom. I'm ready."

Samarie called the meeting to order. There'd be some changes made, she told them. "You should take it slow," Rick Fisher whispered in her ear.

"I'll do nothing of the kind," she said. "The fate of the world is at stake. And," she added, "by the way. You're fired."

Once Fisher had left the room, Samarie passed around the test protocols Walter had developed. "We'll be using these protocols to check GMOs for the pathogen that causes MODS. If it's present, we pull the product. If, as it seems, it's only GMO plants treated with Veresate that are the problem, we just need to prevent any further treatment. Any questions?"

The shareholders shook their heads, not meeting her eyes.

Samarie consulted her tablet. "All right, then. Next, we stop producing Veresate."

Jim Baker, the R&D VP interrupted. "We can't turn it off like a faucet," he protested.

"Why, yes," Samarie looked at his nameplate, "Jim. We can. I checked. And not only do we stop producing it, we make certain all stockpiles are destroyed, worldwide." She offered him a pleasant smile. "You'll take care of that personally, won't you, Jim?"

Baker nodded unhappily, but made a note on his tablet.

Samarie took her time, smiling at each person at the table. "That ought to do for a start," she said. "Later, we'll be looking into the FDA and USDA, exploring why they didn't test our products as thoroughly as they might have...."

"You don't want to go there," Baker warned her.

Samarie turned back to him with a smile. "Oh, but I do, Jim. Would you prefer to not be a part of that...investigation?"

When Baker didn't respond, Samarie continued. "Transparency is our new motto. I've asked Maya King to take the true Veregro story to the public in a way that does not cause any panic, and reassures everyone that we're now working in *their* best interests, not solely the shareholders'. As the principle shareholder, this is in my best interest."

Baker shook his head, then leaned over and whispered in the ear of the man next to him. "She's a chip off the old block."

Samarie offered him another smile. "You're right, Jim. I am a chip off the old block—my mother. Remember that."

Baker nodded.

"So before our final piece of business today, I'd like to make a personal statement." Samarie looked around the table. "I know what you all must be thinking. How can an eighteen-year-old lead this company? She doesn't have the experience and she doesn't have the education. And you're right. Not only that, but I'm a new mom."

The women at the table offered a subdued clap. "Thank you," Samarie said with a smile. "I really owe it to my baby girl to be the best mom I can be. And I'd also like to go to college. Education is so important, and mine's not complete yet. So after giving this much thought, I think it makes sense for me to step down as CEO. This company needs more than I can give right now. I'll stay on as chairman because I want to be involved, but the day-to-day operations belong in the hands of someone with far more than I can offer. For now."

The room perked up and comments from the table could be overheard.

"Very astute decision."

"That shows great maturity."

"Thank God."

Samarie put up a hand to quiet the table. "So for our last piece of business today, I'm proud to introduce you to your new boss, Veregro's new CEO, Ms. Charlie Winters."

ABOUT THE AUTHORS

ROY MANKOVITZ (1941–2011) was a U.S. entrepreneur, rocket scientist, lawyer, inventor, nature-based illness prevention researcher, and author.

Born in New York City, Mankovitz graduated with a degree in engineering science from Columbia University's School of Engineering and Applied Science, and from the University of LaVerne College

 of Law with a juris doctor degree. He was a member of the California and Federal Bars, and a licensed patent attorney.

Mankovitz joined Rocketdyne, then a division of North American Aviation, where he designed and developed engine control systems for Gemini and Apollo spacecraft and the Lunar Descent engine, and developed digital solenoid valve drivers that enabled the control of large valves with a minimum of electrical power.

A few years later, Mankovitz joined the Guidance and Control Division at Jet Propulsion Laboratory, where he authored computer programs and designed control systems for Mars landing vehicles and deep space probes.

In 1968, Mankovitz took a position as director of engineering at a division of Teledyne which produced electromechanical relays, where he developed and patented the first commercially produced solid state relays. Mankovitz also co-founded Chardonnay Corporation, which designed and produced remote-controlled systems that revolutionized the pool and spa industry by automating the operation of pool-spa combinations.

In 1991, Mankovitz joined a start-up company, Gemstar Development Corporation, as its in-house intellectual property counsel, as well as a member of its research and development team. During his tenure, he became a director and officer, and built Gemstar into a patent powerhouse in the field of consumer electronics.

In 1998, Mankovitz founded Patentlab, LLC, devoted to researching, designing, patenting, and licensing his intellectual property. Mankovitz continued his entrepreneurial interests in the field of consumer electronics as co-founder, director, counsel, and chief strategy officer for Web Tuner Corp., a Redmond, Washington, based high-tech startup devoted to the convergence of the internet and television viewing. He also directed his entrepreneurial interests into the field of health by co-founding Berrynol, Inc., to promote his patented technology in the area of topical photo-protective preparations.

In 2005, Mankovitz founded Montecito Wellness, LLC, with his wife, Kathleen Barry, Ph.D., which is dedicated to research in the field of primary illness prevention through technology, products and processes based on nature. The first book he published on the subject is *The Wellness Project – A Rocket Scientist's Blueprint for Health*. Mankovitz published two more books on health and wellness: *Nature's Detox Plan – A Program for Physical and Emotional Detoxification*; and *The Original Diet – The Omnivore's Solution*.

The overall premise of his research is that nature is a template for healthy living, and when married to our evolutionary heritage, optimizes our health, even to the point of overcoming past illnesses.

Roy Mankovitz died on July 10, 2011.

Alan Mankovitz is co-owner of the internet marketing and information service company, Egravitas, which specializes in providing online information about concerts, sports and theatre events.

He previously worked on the creative development of promotional campaigns for clients like Disney and TVG Network, writing much of the marketing and advertising copy. He also has extensive experience with Warner Music Group as a liaison between the creative and manufacturing processes for packaging, advertising, marketing, publicity and promotional materials. He specialized in interpreting artistic visions and bringing them to life on budget and on time. Overseeing production for Warner Bros. Records' entire release schedule, he worked on projects for a wide range of artists including Madonna, Eric Clapton, Red Hot Chili Peppers, R.E.M., Green Day, Joni Mitchell, Neil Young, Tom Petty and many others. He was awarded a Platinum Record by Maverick Records for his assistance on a Madonna release and has spearheaded a number of projects that went on to win Grammy awards for packaging.

Alan holds a BA in film and television production from California State University, Northridge, and in 2004 was asked to produce a documentary on the work of a three-time Grammy award-winning art director. The film has been shown at art and design schools worldwide.

His first published work was writing computer game reviews for Softline and Softalk magazine at the age of fifteen. He considers it a great privilege and delight to complete his late dad's novel.